I0066009

MATRIX THEORY - APPLICATIONS AND THEOREMS

Edited by **Hassan A. Yasser**

Matrix Theory-Applications and Theorems
http://dx.doi.org/10.5772/intechopen.71169
Edited by Hassan A. Yasser

Contributors

Saleem Alashhab, Ivan Kyrchei, Abdul Hamid Ganie, Victor Eduardo Martinez-Luaces, Gabino Torres-Vega, Armando Martínez-Pérez

© **The Editor(s) and the Author(s) 2018**
The rights of the editor(s) and the author(s) have been asserted in accordance with the Copyright, Designs and Patents Act 1988. All rights to the book as a whole are reserved by INTECHOPEN LIMITED. The book as a whole (compilation) cannot be reproduced, distributed or used for commercial or non-commercial purposes without INTECHOPEN LIMITED's written permission. Enquiries concerning the use of the book should be directed to INTECHOPEN LIMITED rights and permissions department (permissions@intechopen.com).
Violations are liable to prosecution under the governing Copyright Law.

(cc) BY

Individual chapters of this publication are distributed under the terms of the Creative Commons Attribution 3.0 Unported License which permits commercial use, distribution and reproduction of the individual chapters, provided the original author(s) and source publication are appropriately acknowledged. If so indicated, certain images may not be included under the Creative Commons license. In such cases users will need to obtain permission from the license holder to reproduce the material. More details and guidelines concerning content reuse and adaptation can be found at http://www.intechopen.com/copyright-policy.html.

Notice

Statements and opinions expressed in the chapters are these of the individual contributors and not necessarily those of the editors or publisher. No responsibility is accepted for the accuracy of information contained in the published chapters. The publisher assumes no responsibility for any damage or injury to persons or property arising out of the use of any materials, instructions, methods or ideas contained in the book.

First published in London, United Kingdom, 2018 by IntechOpen
IntechOpen is the global imprint of INTECHOPEN LIMITED, registered in England and Wales, registration number: 11086078, The Shard, 25th floor, 32 London Bridge Street
London, SE19SG – United Kingdom
Printed in Croatia

British Library Cataloguing-in-Publication Data
A catalogue record for this book is available from the British Library

Additional hard copies can be obtained from orders@intechopen.com

Matrix Theory-Applications and Theorems, Edited by Hassan A. Yasser
p. cm.
Print ISBN 978-1-78923-466-4
Online ISBN 978-1-78923-467-1

We are IntechOpen,
the world's leading publisher of
Open Access books
Built by scientists, for scientists

3,650+
Open access books available

114,000+
International authors and editors

119M+
Downloads

Our authors are among the

151
Countries delivered to

Top 1%
most cited scientists

12.2%
Contributors from top 500 universities

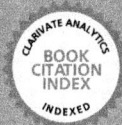

CLARIVATE ANALYTICS
BOOK
CITATION
INDEX
INDEXED

WEB OF SCIENCE™

Selection of our books indexed in the Book Citation Index
in Web of Science™ Core Collection (BKCI)

Interested in publishing with us?
Contact book.department@intechopen.com

Numbers displayed above are based on latest data collected.
For more information visit www.intechopen.com

Meet the editor

Hassan A. Yasser received his BSc and MSc degrees from the science college of the University of Baghdad in 1989 and 1995, respectively. He received his PhD degree from the science college of the University of Basra in 2005. He worked in the department of mathematics for seven years, in the department of computer science for six years, and in the department of physics for eleven years. More than 70 academic papers have been published on various topics of optical communication, image processing, and mathematical physics. He has worked on four books: two books for IntechOpen *Linear Algebra, Theories and Applications* and *Fiber Optic Research*, a book entitled *Analysis of Four-Band Mixing in Optical Semiconductor Amplifiers*, and a book on *Mathematical Physics in Arabic Language*. He supervised 20 graduate students for a number of Iraqi universities.

Contents

Preface

Matrices have applications in a huge number of scientific fields. In physics, they are used to study physical phenomena, such as the motion of rigid bodies. In computer graphics, they are used to manipulate 3D models and project them onto a 2D screen. In probability theory and statistics, stochastic matrices are used to describe sets of probabilities. Matrices calculus may be used in economics to describe systems of economic relationships. A main part of numerical analysis focuses on the development of efficient algorithms for matrix computations, a subject that is centuries old and is today an expanding area of research. Matrix decomposition methods simplify computations, both theoretically and practically. Algorithms that are tailored to particular matrix structures, such as sparse matrices and near-diagonal matrices, expedite computations in the finite element method and other computations. Infinite matrices are present in planetary theory and atomic theory. They are the matrices representing the derivative operators, which act on the Taylor series of a function. This new book reviews current research, including applications of matrices and spaces, as well as other characteristics.

The book is divided into two sections. The first section (Chapters 1 and 2) discusses the application of matrices that has become an area of academic research and of great importance in many scientific fields. In Chapter 1, within the framework of the theory of row/column determinants, the determinantal representations (analogs of Cramer's rule) of a partial solution to the system of two-sided quaternion matrix equations, $A_1XB_1=C_1$, $A_2XB_2=C_2$ are analyzed. It also gives Cramer's rules for its special cases with one-sided equations and considers the two systems with the first equation $A_1X=C_1$ and $XB_1=C_1$, respectively, and with an unchanging second equation. Cramer's rules for special cases when two equations are one sided, to wit, the system of equations $A_1X=C_1$, $XB_2=C_2$ and the system of the equations $A_1X=C_1$, $A_2X=C_2$, are studied as well. Chapter 2 introduces and studies a matrix that has the exponential function as one of its eigenvectors and realizes that this matrix represents finite difference derivation of vectors on a partition. This matrix leads to new expressions for finite difference derivatives, which are exact for the exponential function. A number of properties of this matrix, induced derivatives, and its inverse are also found. In addition, the expression for the derivative of a product, a ratio, and the inverse of vectors plus the equivalent of the summation by parts theorem of continuous functions are also described. This matrix could be of interest to discrete quantum mechanics theory.

The second section (Chapters 3 to 5) comprises three chapters discussing spaces and linear systems. In Chapter 3, mixing problems are considered since they always lead to linear ODE systems, and the corresponding associated matrices have different structures that deserve to be studied in depth. This structure depends on whether there is recirculation of fluids and if the system is open or closed, among other characteristics such as the number of tanks and their internal connections. Several statements regarding matrix eigenvalues are analyzed for different structures and also some questions and conjectures are posed. Finally, qualitative remarks about differential equation system solutions and their stability or asymptotical stability are included. In Chapter 4, special compound ′ magic squares are considered and a -dimensional subspace of the nullspace of the ′ squares is determined. All vectors in the subspaces possess the property that the sum of all entries of each vector equals zero. In Chapter 5, a new type of regular matrix generated by Fibonacci numbers is introduced and we shall investigate its various topological properties. The concept of mathematical regularity in terms of Fibonacci numbers and phyllotaxy have been discussed.

Prof. Dr. Hassan A. Yasser
Thi-Qar University
Science College
Physics Department
Iraq

Applications of Matrices

Cramer's Rules for the System of Two-Sided Matrix Equations and of Its Special Cases

Ivan I. Kyrchei

Additional information is available at the end of the chapter

http://dx.doi.org/10.5772/intechopen.74105

Abstract

Within the framework of the theory of row-column determinants previously introduced by the author, we get determinantal representations (analogs of Cramer's rule) of a partial solution to the system of two-sided quaternion matrix equations $A_1XB_1=C_1$, $A_2XB_2=C_2$. We also give Cramer's rules for its special cases when the first equation be one-sided. Namely, we consider the two systems with the first equation $A_1X=C_1$ and $XB_1=C_1$, respectively, and with an unchanging second equation. Cramer's rules for special cases when two equations are one-sided, namely the system of the equations $A_1X=C_1$, $XB_2=C_2$, and the system of the equations $A_1X=C_1$, $A_2X=C_2$ are studied as well. Since the Moore-Penrose inverse is a necessary tool to solve matrix equations, we use its determinantal representations previously obtained by the author in terms of row-column determinants as well.

Keywords: Moore-Penrose inverse, quaternion matrix, Cramer rule, system matrix equations

2000 AMS subject classifications: 15A15, 16 W10

1. Introduction

The study of matrix equations and systems of matrix equations is an active research topic in matrix theory and its applications. The system of classical two-sided matrix equations

$$\begin{cases} A_1XB_1 = C_1, \\ A_2XB_2 = C_2. \end{cases} \tag{1}$$

over the complex field, a principle domain, and the quaternion skew field has been studied by many authors (see, e.g. [1–7]). Mitra [1] gives necessary and sufficient conditions of the system

© 2018 The Author(s). Licensee IntechOpen. This chapter is distributed under the terms of the Creative Commons Attribution License (http://creativecommons.org/licenses/by/3.0), which permits unrestricted use, distribution, and reproduction in any medium, provided the original work is properly cited. (cc) BY

IntechOpen

(1) over the complex field and the expression for its general solution. Navarra et al. [6] derived a new necessary and sufficient condition for the existence and a new representation of (1) over the complex field and used the results to give a simple representation. Wang [7] considers the system (1) over the quaternion skew field and gets its solvability conditions and a representation of a general solution.

Throughout the chapter, we denote the real number field by $\mathbb{R}$, the set of all $m \times n$ matrices over the quaternion algebra

$$\mathbb{H} = \left\{ a_0 + a_1 i + a_2 j + a_3 k \,|\, i^2 = j^2 = k^2 = -1, a_0, a_1, a_2, a_3 \in \mathbb{R} \right\}$$

by $\mathbb{H}^{m \times n}$ and by $\mathbb{H}_r^{m \times n}$, and the set of matrices over $\mathbb{H}$ with a rank r. For $\mathbf{A} \in \mathbb{H}^{n \times m}$, the symbols $\mathbf{A}^*$ stands for the conjugate transpose (Hermitian adjoint) matrix of $\mathbf{A}$. The matrix $\mathbf{A} = (a_{ij}) \in \mathbb{H}^{n \times n}$ is Hermitian if $\mathbf{A}^* = \mathbf{A}$.

Generalized inverses are useful tools used to solve matrix equations. The definitions of the Moore-Penrose inverse matrix have been extended to quaternion matrices as follows. The Moore-Penrose inverse of $\mathbf{A} \in \mathbb{H}^{m \times n}$, denoted by $\mathbf{A}^\dagger$, is the unique matrix $\mathbf{X} \in \mathbb{H}^{n \times m}$ satisfying (1) $\mathbf{AXA} = \mathbf{A}$, (2) $\mathbf{XAX} = \mathbf{X}$, (3) $(\mathbf{AX})^* = \mathbf{AX}$, and (4) $(\mathbf{XA})^* = \mathbf{XA}$.

The determinantal representation of the usual inverse is the matrix with the cofactors in the entries which suggests a direct method of finding of inverse and makes it applicable through Cramer's rule to systems of linear equations. The same is desirable for the generalized inverses. But there is not so unambiguous even for complex or real generalized inverses. Therefore, there are various determinantal representations of generalized inverses because of looking for their more applicable explicit expressions (see, e.g. [8]). Through the noncommutativity of the quaternion algebra, difficulties arise already in determining the quaternion determinant (see, e.g. [9–16]).

The understanding of the problem for determinantal representation of an inverse matrix as well as generalized inverses only now begins to be decided due to the theory of column-row determinants introduced in [17, 18]. Within the framework of the theory of column-row determinants, determinantal representations of various kinds of generalized inverses and (generalized inverses) solutions of quaternion matrix equations have been derived by the author (see, e.g. [19–25]) and by other reseachers (see, e.g. [26–29]).

The main goals of the chapter are deriving determinantal representations (analogs of the classical Cramer rule) of general solutions of the system (1) and its simpler cases over the quaternion skew field.

The chapter is organized as follows. In Section 2, we start with preliminaries introducing row-column determinants and determinantal representations of the Moore-Penrose and Cramer's rule of the quaternion matrix equations, $\mathbf{AXB} = \mathbf{C}$. Determinantal representations of a partial solution (an analog of Cramer's rule) of the system (1) are derived in Section 3. In Section 4, we give Cramer's rules to special cases of (1) with 1 and 2 one-sided equations. Finally, the conclusion is drawn in Section 5.

2. Preliminaries

For $\mathbf{A} = (a_{ij}) \in M(n, \mathbb{H})$, we define n row determinants and n column determinants as follows. Suppose S_n is the symmetric group on the set $I_n = \{1, ..., n\}$.

Definition 2.1. *The ith row determinant of* $\mathbf{A} \in \mathbb{H}^{n \times m}$ *is defined for all* $i = 1, ..., n$ *by putting*

$$\operatorname{rdet}_i \mathbf{A} = \sum_{\sigma \in S_n} (-1)^{n-r} \left(a_{i i_{k_1}} a_{i_{k_1} i_{k_1}+1} \ldots a_{i_{k_1}+l_1 i} \right) \ldots \left(a_{i_{k_r} i_{k_r}+1} \ldots a_{i_{k_r}+l_r i_{k_r}} \right),$$

$$\sigma = (i i_{k_1} i_{k_1}+1 \ldots i_{k_1}+l_1)(i_{k_2} i_{k_2}+1 \ldots i_{k_2}+l_2) \ldots (i_{k_r} i_{k_r}+1 \ldots i_{k_r}+l_r),$$

with conditions $i_{k_2} < i_{k_3} < ... < i_{k_r}$ and $i_{k_t} < i_{k_t+s}$ for all $t = 2, ..., r$ and all $s = 1, ..., l_t$.

Definition 2.2. *The jth column determinant of* $\mathbf{A} \in \mathbb{H}^{n \times m}$ *is defined for all* $j = 1, ..., n$ *by putting*

$$\operatorname{cdet}_j \mathbf{A} = \sum_{\tau \in S_n} (-1)^{n-r} \left(a_{j_{k_r} j_{k_r}+l_r} \ldots a_{j_{k_r}+1 i_{k_r}} \right) \ldots \left(a_{j j_{k_1}+l_1} \ldots a_{j_{k_1}+1 j_{k_1}} a_{j_{k_1} j} \right),$$

$$\tau = \left(j_{k_r}+l_r \ldots j_{k_r}+1 j_{k_r} \right) \ldots \left(j_{k_2}+l_2 \ldots j_{k_2}+1 j_{k_2} \right) \left(j_{k_1}+l_1 \ldots j_{k_1}+1 j_{k_1} j \right),$$

with conditions, $j_{k_2} < j_{k_3} < ... < j_{k_r}$ and $j_{k_t} < j_{k_t+s}$ for $t = 2, ..., r$ and $s = 1, ..., l_t$.

Since $\operatorname{rdet}_1 \mathbf{A} = \cdots = \operatorname{rdet}_n \mathbf{A} = \operatorname{cdet}_1 \mathbf{A} = \cdots = \operatorname{cdet}_n \mathbf{A} \in \mathbb{R}$ for Hermitian $\mathbf{A} \in \mathbb{H}^{n \times n}$, then we can define the determinant of a Hermitian matrix $\mathbf{A}$ by putting, $\det \mathbf{A} := \operatorname{rdet}_i \mathbf{A} = \operatorname{cdet}_i \mathbf{A}$, for all $i = 1, ..., n$. The determinant of a Hermitian matrix has properties similar to a usual determinant. They are completely explored in [17, 18] by its row and column determinants. In particular, within the framework of the theory of the column-row determinants, the determinantal representations of the inverse matrix over $\mathbb{H}$ by analogs of the classical adjoint matrix and Cramer's rule for quaternionic systems of linear equations have been derived. Further, we consider the determinantal representations of the Moore-Penrose inverse.

We shall use the following notations. Let $\alpha := \{\alpha_1, ..., \alpha_k\} \subseteq \{1, ..., m\}$ and $\beta := \{\beta_1, ..., \beta_k\} \subseteq \{1, ..., n\}$ be subsets of the order $1 \leq k \leq \min\{m, n\}$. $\mathbf{A}_\beta^\alpha$ denotes the submatrix of $\mathbf{A} \in \mathbb{H}^{n \times m}$ determined by the rows indexed by α and the columns indexed by β. Then, $\mathbf{A}_\alpha^\alpha$ denotes the principal submatrix determined by the rows and columns indexed by α. If $\mathbf{A} \in \mathbb{H}^{n \times n}$ is Hermitian, then $|\mathbf{A}|_\alpha^\alpha$ is the corresponding principal minor of $\det \mathbf{A}$. For $1 \leq k \leq n$, the collection of strictly increasing sequences of k integers chosen from $\{1, ..., n\}$ is denoted by $L_{k,n} := \{\alpha : \alpha = (\alpha_1, ..., \alpha_k), 1 \leq \alpha_1 \leq ... \leq \alpha_k \leq n\}$. For fixed $i \in \alpha$ and $j \in \beta$, let $I_{r,m}\{i\} := \{\alpha : \alpha \in L_{r,m}, i \in \alpha\}$, $J_{r,n}\{j\} := \{\beta : \beta \in L_{r,n}, j \in \beta\}$.

Let $\mathbf{a}_{.j}$ be the jth column and $\mathbf{a}_{i.}$ be the ith row of $\mathbf{A}$. Suppose $\mathbf{A}_{.j}(\mathbf{b})$ denotes the matrix obtained from $\mathbf{A}$ by replacing its jth column with the column $\mathbf{b}$, then $\mathbf{A}_{i.}(\mathbf{b})$ denotes the matrix obtained from $\mathbf{A}$ by replacing its ith row with the row $\mathbf{b}$. $\mathbf{a}_{.j}^{*}$ and $\mathbf{a}_{i.}^{*}$ denote the jth column and the ith row of $\mathbf{A}^{*}$, respectively.

The following theorem gives determinantal representations of the Moore-Penrose inverse over the quaternion skew field $\mathbb{H}$.

Theorem 2.1. [19] *If* $\mathbf{A} \in \mathbb{H}_r^{m \times n}$, *then the Moore-Penrose inverse* $\mathbf{A}^{\dagger} = \left(a_{ij}^{\dagger} \right) \in \mathbb{H}^{n \times m}$ *possesses the following determinantal representations:*

$$a_{ij}^{\dagger} = \frac{\sum_{\beta \in J_{r,n}\{i\}} \operatorname{cdet}_i \left((\mathbf{A}^*\mathbf{A})_{.i} \left(\mathbf{a}_{.j}^* \right) \right)_{\beta}^{\beta}}{\sum_{\beta \in J_{r,n}} |\mathbf{A}^*\mathbf{A}|_{\beta}^{\beta}}, \tag{2}$$

or

$$a_{ij}^{\dagger} = \frac{\sum_{\alpha \in I_{r,m}\{j\}} \operatorname{rdet}_j \left((\mathbf{A}\mathbf{A}^*)_{j.} \left(\mathbf{a}_{i.}^* \right) \right)_{\alpha}^{\alpha}}{\sum_{\alpha \in I_{r,m}} |\mathbf{A}\mathbf{A}^*|_{\alpha}^{\alpha}}. \tag{3}$$

Remark 2.1. *Note that for an arbitrary full-rank matrix,* $\mathbf{A} \in \mathbb{H}_r^{m \times n}$, *a column-vector* $\mathbf{d}_{.j}$, *and a row-vector* $\mathbf{d}_{i.}$ *with appropriate sizes, respectively, we put*

$$\operatorname{cdet}_i \left((\mathbf{A}^*\mathbf{A})_{.i} (\mathbf{d}_{.j}) \right) = \sum_{\beta \in J_{n,n}\{i\}} \operatorname{cdet}_i \left((\mathbf{A}^*\mathbf{A})_{.i} (\mathbf{d}_{.j}) \right)_{\beta}^{\beta}, \quad \det(\mathbf{A}^*\mathbf{A}) = \sum_{\beta \in J_{n,n}} |\mathbf{A}^*\mathbf{A}|_{\beta}^{\beta} \text{ when } r = n,$$

$$\operatorname{rdet}_j \left((\mathbf{A}\mathbf{A}^*)_{j.} (\mathbf{d}_{i.}) \right) = \sum_{\alpha \in I_{m,m}\{j\}} \operatorname{rdet}_j \left((\mathbf{A}\mathbf{A}^*)_{j.} (\mathbf{d}_{i.}) \right)_{\alpha}^{\alpha}, \quad \det(\mathbf{A}\mathbf{A}^*) = \sum_{\alpha \in I_{m,m}} |\mathbf{A}\mathbf{A}^*|_{\alpha}^{\alpha} \text{ when } r = m.$$

Furthermore, $\mathbf{P}_A = \mathbf{A}^{\dagger}\mathbf{A}$, $\mathbf{Q}_A = \mathbf{A}\mathbf{A}^{\dagger}$, $\mathbf{L}_A = \mathbf{I} - \mathbf{A}^{\dagger}\mathbf{A}$, and $\mathbf{R}_A := \mathbf{I} - \mathbf{A}\mathbf{A}^{\dagger}$ stand for some orthogonal projectors induced from $\mathbf{A}$.

Theorem 2.2. [30] *Let* $\mathbf{A} \in \mathbb{H}^{m \times n}$, $\mathbf{B} \in \mathbb{H}^{r \times s}$, *and* $\mathbf{C} \in \mathbb{H}^{m \times s}$ *be known and* $\mathbf{X} \in \mathbb{H}^{n \times r}$ *be unknown. Then, the matrix equation*

$$\mathbf{A}\mathbf{X}\mathbf{B} = \mathbf{C} \tag{4}$$

is consistent if and only if $\mathbf{A}\mathbf{A}^{\dagger}\mathbf{C}\mathbf{B}\mathbf{B}^{\dagger} = \mathbf{C}$. *In this case, its general solution can be expressed as*

$$\mathbf{X} = \mathbf{A}^{\dagger}\mathbf{C}\mathbf{B}^{\dagger} + \mathbf{L}_A\mathbf{V} + \mathbf{W}\mathbf{R}_B, \tag{5}$$

where V and W are arbitrary matrices over $\mathbb{H}$ *with appropriate dimensions.*

The partial solution, $\mathbf{X}^0 = \mathbf{A}^{\dagger}\mathbf{C}\mathbf{B}^{\dagger}$, of (4) possesses the following determinantal representations.

Theorem 2.3. [20] *Let* $\mathbf{A} \in \mathbb{H}_{r_1}^{m \times n}$ *and* $\mathbf{B} \in \mathbb{H}_{r_2}^{r \times s}$. *Then,* $\mathbf{X}^0 = \left(x_{ij}^0 \right) \in \mathbb{H}^{n \times r}$ *has determinantal representations,*

$$x_{ij} = \frac{\sum_{\beta \in J_{r_1,n}\{i\}} \operatorname{cdet}_i\left((\mathbf{A}^*\mathbf{A})_{\cdot i}\left(\mathbf{d}_{\cdot j}^B\right)\right)_\beta^\beta}{\sum_{\beta \in J_{r_1,n}} |\mathbf{A}^*\mathbf{A}|_\beta^\beta \sum_{\alpha \in I_{r_2,r}} |\mathbf{BB}^*|_\alpha^\alpha},$$

or

$$x_{ij} = \frac{\sum_{\alpha \in I_{r_2,r}\{j\}} \operatorname{rdet}_j\left((\mathbf{BB}^*)_{j\cdot}\left(\mathbf{d}_{i\cdot}^A\right)\right)_\alpha^\alpha}{\sum_{\beta \in J_{r_1,n}} |\mathbf{A}^*\mathbf{A}|_\beta^\beta \sum_{\alpha \in I_{r_2,r}} |\mathbf{BB}^*|_\alpha^\alpha},$$

where

$$\mathbf{d}_{\cdot j}^B = \left[\sum_{\alpha \in I_{r_2,r}\{j\}} \operatorname{rdet}_j\left((\mathbf{BB}^*)_{j\cdot}(\tilde{\mathbf{c}}_{k\cdot})\right)_\alpha^\alpha \right] \in \mathbb{H}^{n \times 1}, \;\; k = 1, \ldots, n,$$

$$\mathbf{d}_{i\cdot}^A = \left[\sum_{\beta \in J_{r_1,n}\{i\}} \operatorname{cdet}_i\left((\mathbf{A}^*\mathbf{A})_{\cdot i}(\tilde{\mathbf{c}}_{\cdot l})\right)_\beta^\beta \right] \in \mathbb{H}^{1 \times r}, \;\; l = 1, \ldots, r,$$

are the column vector and the row vector, respectively. $\tilde{\mathbf{c}}_{i\cdot}$ *and* $\tilde{\mathbf{c}}_{\cdot j}$ *are the ith row and the jth column of* $\tilde{\mathbf{C}} = \mathbf{A}^*\mathbf{C}\mathbf{B}^*$.

3. Determinantal representations of a partial solution to the system (1)

Lemma 3.1. [7] *Let* $\mathbf{A}_1 \in \mathbb{H}^{m \times n}$, $\mathbf{B}_1 \in \mathbb{H}^{r \times s}$, $\mathbf{C}_1 \in \mathbb{H}^{m \times s}$, $\mathbf{A}_2 \in \mathbb{H}^{k \times n}$, $\mathbf{B}_2 \in \mathbb{H}^{r \times p}$, *and* $\mathbf{C}_2 \in \mathbb{H}^{k \times p}$ *be given and* $\mathbf{X} \in \mathbb{H}^{n \times r}$ *is to be determined. Put* $\mathbf{H} = \mathbf{A}_2 \mathbf{L}_{A_1}$, $\mathbf{N} = \mathbf{R}_{B_1} \mathbf{B}_2$, $\mathbf{T} = \mathbf{R}_H \mathbf{A}_2$, *and* $\mathbf{F} = \mathbf{B}_2 \mathbf{L}_N$. *Then, the system (1) is consistent if and only if*

$$\mathbf{A}_i \mathbf{A}_i^\dagger \mathbf{C}_i \mathbf{B}_i^\dagger \mathbf{B}_i = \mathbf{C}_i, \quad i = 1, 2; \tag{6}$$

$$\mathbf{T}\left[\mathbf{A}_2^\dagger \mathbf{X} \mathbf{B}_2^\dagger - \mathbf{A}_1^\dagger \mathbf{C}_1 \mathbf{B}_1^\dagger\right]\mathbf{F} = 0. \tag{7}$$

In that case, the general solution of (1) can be expressed as the following,

$$\begin{aligned}
\mathbf{X} = \;& \mathbf{A}_1^\dagger \mathbf{C}_1 \mathbf{B}_1^\dagger + \mathbf{L}_{A_1} \mathbf{H}^\dagger \mathbf{A}_2 \mathbf{L}_T \left(\mathbf{A}_2^\dagger \mathbf{C}_2 \mathbf{B}_2^\dagger - \mathbf{A}_1^\dagger \mathbf{C}_1 \mathbf{B}_1^\dagger\right)\mathbf{B}_2 \mathbf{B}_2^\dagger + \mathbf{T}^\dagger \mathbf{T}\left(\mathbf{A}_2^\dagger \mathbf{C}_2 \mathbf{B}_2^\dagger - \mathbf{A}_1^\dagger \mathbf{C}_1 \mathbf{B}_1^\dagger\right)\mathbf{B}_2 \mathbf{N}^\dagger \mathbf{R}_{B_1} \\
& + \mathbf{L}_{A_1}\left(\mathbf{Z} - \mathbf{H}^\dagger \mathbf{H} \mathbf{Z} \mathbf{B}_2 \mathbf{B}_2^\dagger\right) - \mathbf{L}_{A_1} \mathbf{H}^\dagger \mathbf{A}_2 \mathbf{L}_T \mathbf{W} \mathbf{N} \mathbf{B}_2^\dagger + \left(\mathbf{W} - \mathbf{T}^\dagger \mathbf{T} \mathbf{W} \mathbf{N} \mathbf{N}^\dagger\right) \times \mathbf{R}_{B_1},
\end{aligned} \tag{8}$$

where $\mathbf{Z}$ *and* $\mathbf{W}$ *are the arbitrary matrices over* $\mathbb{H}$ *with compatible dimensions.*

Some simplification of (8) can be derived due to the quaternionic analog of the following proposition.

Lemma 3.2. [32] *If* $\mathbf{A} \in \mathbb{H}^{n \times n}$ *is Hermitian and idempotent, then the following equation holds for any matrix* $\mathbf{B} \in \mathbb{H}^{m \times n}$,

$$\mathbf{A}(\mathbf{BA})^\dagger = (\mathbf{BA})^\dagger. \tag{9}$$

It is evident that if $\mathbf{A} \in \mathbb{H}^{n \times n}$ is Hermitian and idempotent, then the following equation is true as well,

$$(\mathbf{AB})^\dagger \mathbf{A} = (\mathbf{AB})^\dagger. \tag{10}$$

Since $\mathbf{L}_{A_1}$, $\mathbf{R}_{B_1}$, and $\mathbf{R}_H$ are projectors, then using (9) and (10), we have, respectively,

$$\begin{aligned}
\mathbf{L}_{A_1}\mathbf{H}^\dagger &= \mathbf{L}_{A_1}(\mathbf{A}_2\mathbf{L}_{A_1})^\dagger = (\mathbf{A}_2\mathbf{L}_{A_1})^\dagger = \mathbf{H}^\dagger, \\
\mathbf{N}^\dagger\mathbf{R}_{B_1} &= (\mathbf{R}_{B_1}\mathbf{B}_2)^\dagger\mathbf{R}_{B_1} = (\mathbf{R}_{B_1}\mathbf{B}_2)^\dagger = \mathbf{N}^\dagger, \\
\mathbf{T}^\dagger\mathbf{T} &= (\mathbf{R}_H\mathbf{A}_2)^\dagger\mathbf{R}_H\mathbf{A}_2 = (\mathbf{R}_H\mathbf{A}_2)^\dagger\mathbf{A}_2 = \mathbf{T}^\dagger\mathbf{A}_2, \\
\mathbf{L}_T &= \mathbf{I} - \mathbf{T}^\dagger\mathbf{T} = \mathbf{I} - \mathbf{T}^\dagger\mathbf{A}_2.
\end{aligned} \tag{11}$$

Using (11) and (6), we obtain the following expression of (8),

$$\begin{aligned}
\mathbf{X} &= \mathbf{A}_1^\dagger\mathbf{C}_1\mathbf{B}_1^\dagger + \mathbf{H}^\dagger\mathbf{A}_2(\mathbf{I} - \mathbf{T}^\dagger\mathbf{A}_2)(\mathbf{A}_2^\dagger\mathbf{C}_2\mathbf{B}_2^\dagger - \mathbf{A}_1^\dagger\mathbf{C}_1\mathbf{B}_1^\dagger)\mathbf{B}_2\mathbf{B}_2^\dagger \\
&\quad + \mathbf{T}^\dagger\mathbf{A}_2(\mathbf{A}_2^\dagger\mathbf{C}_2\mathbf{B}_2^\dagger - \mathbf{A}_1^\dagger\mathbf{C}_1\mathbf{B}_1^\dagger)\mathbf{B}_2\mathbf{N}^\dagger + \mathbf{L}_{A_1}(\mathbf{Z} - \mathbf{H}^\dagger\mathbf{HZB}_2\mathbf{B}_2^\dagger) - \mathbf{H}^\dagger\mathbf{A}_2\mathbf{L}_T\mathbf{WNB}_2^\dagger \\
&\quad + (\mathbf{W} - \mathbf{T}^\dagger\mathbf{TWNN}^\dagger)\mathbf{R}_{B_1} = \mathbf{A}_1^\dagger\mathbf{C}_1\mathbf{B}_1^\dagger + \mathbf{H}^\dagger\mathbf{C}_2\mathbf{B}_2^\dagger + \mathbf{H}^\dagger(\mathbf{A}_2\mathbf{T}^\dagger - \mathbf{I})\mathbf{A}_2\mathbf{A}_1^\dagger\mathbf{C}_1\mathbf{B}_1^\dagger\mathbf{Q}_{B_2} \\
&\quad - \mathbf{H}^\dagger\mathbf{A}_2\mathbf{T}^\dagger\mathbf{C}_2\mathbf{B}_2^\dagger + \mathbf{T}^\dagger\mathbf{C}_2\mathbf{N}^\dagger - \mathbf{T}^\dagger\mathbf{A}_2\mathbf{A}_1^\dagger\mathbf{C}_1\mathbf{B}_1^\dagger\mathbf{B}_2\mathbf{N}^\dagger + \mathbf{L}_{A_1}(\mathbf{Z} - \mathbf{H}^\dagger\mathbf{HZB}_2\mathbf{B}_2^\dagger) \\
&\quad - \mathbf{H}^\dagger\mathbf{A}_2\mathbf{L}_T\mathbf{WNB}_2^\dagger + (\mathbf{W} - \mathbf{T}^\dagger\mathbf{TWNN}^\dagger)\mathbf{R}_{B_1}.
\end{aligned} \tag{12}$$

By putting $\mathbf{Z}_1 = \mathbf{W}_1 = 0$ in (12), the partial solution of (8) can be derived,

$$\begin{aligned}
\mathbf{X}_0 &= \mathbf{A}_1^\dagger\mathbf{C}_1\mathbf{B}_1^\dagger + \mathbf{H}^\dagger\mathbf{C}_2\mathbf{B}_2^\dagger + \mathbf{T}^\dagger\mathbf{C}_2\mathbf{N}^\dagger + \mathbf{H}^\dagger\mathbf{A}_2\mathbf{T}^\dagger\mathbf{A}_2\mathbf{A}_1^\dagger\mathbf{C}_1\mathbf{B}_1^\dagger\mathbf{Q}_{B_2} \\
&\quad - \mathbf{H}^\dagger\mathbf{A}_2\mathbf{A}_1^\dagger\mathbf{C}_1\mathbf{B}_1^\dagger\mathbf{Q}_{B_2} - \mathbf{H}^\dagger\mathbf{A}_2\mathbf{T}^\dagger\mathbf{C}_2\mathbf{B}_2^\dagger - \mathbf{T}^\dagger\mathbf{A}_2\mathbf{A}_1^\dagger\mathbf{C}_1\mathbf{B}_1^\dagger\mathbf{B}_2\mathbf{N}^\dagger.
\end{aligned} \tag{13}$$

Further we give determinantal representations of (13). Let $\mathbf{A}_1 = \left(a_{ij}^{(1)}\right) \in \mathbb{H}_{r_1}^{m \times n}$, $\mathbf{B}_1 = \left(b_{ij}^{(1)}\right)$ $\in \mathbb{H}_{r_2}^{r \times s}$, $\mathbf{A}_2 = \left(a_{ij}^{(2)}\right) \in \mathbb{H}_{r_3}^{k \times n}$, $\mathbf{B}_2 = \left(b_{ij}^{(2)}\right) \in \mathbb{H}_{r_4}^{r \times p}$, $\mathbf{C}_1 = \left(c_{ij}^{(1)}\right) \in \mathbb{H}^{m \times s}$, and $\mathbf{C}_2 = \left(c_{ij}^{(2)}\right) \in \mathbb{H}^{k \times p}$, and there exist $\mathbf{A}_1^\dagger = \left(a_{ij}^{(1),\,\dagger}\right) \in \mathbb{H}^{n \times m}$, $\mathbf{B}_2^\dagger = \left(b_{ij}^{(2),\,\dagger}\right) \in \mathbb{H}^{p \times r}$, $\mathbf{H}^\dagger = \left(h_{ij}^\dagger\right) \in \mathbb{H}^{n \times k}$, $\mathbf{N}^\dagger = \left(n_{ij}^\dagger\right) \in \mathbb{H}^{p \times r}$, and $\mathbf{T}^\dagger = \left(t_{ij}^\dagger\right) \in \mathbb{H}^{n \times k}$. Let $\operatorname{rank}\mathbf{H} = \min\{\operatorname{rank}\mathbf{A}_2, \operatorname{rank}\mathbf{L}_{A_1}\} = r_5$, $\operatorname{rank}\mathbf{N} = \min\{\operatorname{rank}\mathbf{B}_2, \operatorname{rank}\mathbf{R}_{B_1}\} = r_6$, and $\operatorname{rank}\mathbf{T} = \min\{\operatorname{rank}\mathbf{A}_2, \operatorname{rank}\mathbf{R}_H\} = r_7$. Consider each term of (13) separately.

(i) By Theorem 2.3 for the first term, x_{ij}^{01}, of (13), we have

$$x_{ij}^{01} = \frac{\sum_{\beta \in J_{r_1,n}\{i\}} \operatorname{cdet}_i\left(\left(\mathbf{A}_1^*\mathbf{A}_1\right)_{.i}\left(\mathbf{d}_{.j}^{B_1}\right)\right)_{\beta}^{\beta}}{\sum_{\beta \in J_{r_1,n}} \left|\mathbf{A}_1^*\mathbf{A}_1\right|_{\beta}^{\beta} \sum_{\alpha \in I_{r_2,p}} \left|\mathbf{B}_1\mathbf{B}_1^*\right|_{\alpha}^{\alpha}},$$

(14)

or

$$x_{ij}^{01} = \frac{\sum_{\alpha \in I_{r_2,q}\{j\}} \operatorname{rdet}_j\left(\left(\mathbf{B}_1\mathbf{B}_1^*\right)_{j.}\left(\mathbf{d}_{i.}^{A_1}\right)\right)_{\alpha}^{\alpha}}{\sum_{\beta \in J_{r_1,p}} \left|\mathbf{A}_1^*\mathbf{A}_1\right|_{\beta}^{\beta} \sum_{\alpha \in I_{r_2,q}} \left|\mathbf{B}_1\mathbf{B}_1^*\right|_{\alpha}^{\alpha}},$$

(15)

where

$$\mathbf{d}_{.j}^{B_1} = \left[\sum_{\alpha \in I_{r_2,p}\{j\}} \operatorname{rdet}_j\left(\left(\mathbf{B}_1\mathbf{B}_1^*\right)_{j.}\left(\tilde{\mathbf{c}}_{q.}^{(1)}\right)\right)_{\alpha}^{\alpha}\right] \in \mathbb{H}^{n \times 1}, \quad q = 1, \ldots, n,$$

$$\mathbf{d}_{i.}^{A_1} = \left[\sum_{\beta \in J_{r_1,n}\{i\}} \operatorname{cdet}_i\left(\left(\mathbf{A}_1^*\mathbf{A}_1\right)_{.i}\left(\tilde{\mathbf{c}}_{.l}^{(1)}\right)\right)_{\beta}^{\beta}\right] \in \mathbb{H}^{1 \times r}, \quad l = 1, \ldots, r,$$

are the column vector and the row vector, respectively. $\tilde{\mathbf{c}}_{q.}^{(1)}$ and $\tilde{\mathbf{c}}_{.l}^{(1)}$ are the qth row and the lth column of $\tilde{\mathbf{C}}_1 = \mathbf{A}_1^*\mathbf{C}_1\mathbf{B}_1^*$.

(ii) Similarly, for the second term of (13), we have

$$x_{ij}^{02} = \frac{\sum_{\beta \in J_{r_5,n}\{i\}} \operatorname{cdet}_i\left(\left(\mathbf{H}^*\mathbf{H}\right)_{.i}\left(\mathbf{d}_{.j}^{B_2}\right)\right)_{\beta}^{\beta}}{\sum_{\beta \in J_{r_5,n}} \left|\mathbf{H}^*\mathbf{H}\right|_{\beta}^{\beta} \sum_{\alpha \in I_{r_4,r}} \left|\mathbf{B}_2\mathbf{B}_2^*\right|_{\alpha}^{\alpha}},$$

(16)

or

$$x_{ij}^{02} = \frac{\sum_{\alpha \in I_{r_4,r}\{j\}} \operatorname{rdet}_j\left(\left(\mathbf{B}_2\mathbf{B}_2^*\right)_{j.}\left(\mathbf{d}_{i.}^{H}\right)\right)_{\alpha}^{\alpha}}{\sum_{\beta \in J_{r_5,n}} \left|\mathbf{H}^*\mathbf{H}\right|_{\beta}^{\beta} \sum_{\alpha \in I_{r_4,r}} \left|\mathbf{B}_2\mathbf{B}_2^*\right|_{\alpha}^{\alpha}},$$

(17)

where

$$\mathbf{d}_{.j}^{B_2} = \left[\sum_{\alpha \in I_{r_4,r}\{j\}} \operatorname{rdet}_j\left(\left(\mathbf{B}_2\mathbf{B}_2^*\right)_{j.}\left(\tilde{\mathbf{c}}_{q.}^{(2)}\right)\right)_{\alpha}^{\alpha}\right] \in \mathbb{H}^{n \times 1}, \quad q = 1, \ldots, n,$$

$$\mathbf{d}_{i.}^{H} = \left[\sum_{\beta \in J_{r_5,n}\{i\}} \operatorname{cdet}_i\left(\left(\mathbf{H}^*\mathbf{H}\right)_{.i}\left(\tilde{\mathbf{c}}_{.l}^{(2)}\right)\right)_{\beta}^{\beta}\right] \in \mathbb{H}^{1 \times r}, \quad l = 1, \ldots, r,$$

are the column vector and the row vector, respectively. $\tilde{\mathbf{c}}_{q.}^{(2)}$ and $\tilde{\mathbf{c}}_{.l}^{(2)}$ are the qth row and the lth column of $\tilde{\mathbf{C}}_2 = \mathbf{H}^*\mathbf{C}_2\mathbf{B}_2^*$. Note that $\mathbf{H}^*\mathbf{H} = (\mathbf{A}_2\mathbf{L}_{A_1})^*\mathbf{A}_2\mathbf{L}_{A_1} = \mathbf{L}_{A_1}\mathbf{A}_2^*\mathbf{A}_2\mathbf{L}_{A_1}$.

(iii) The third term of (13) can be obtained by Theorem 2.3 as well. Then

$$
x_{ij}^{03} = \frac{\sum_{\beta \in J_{r_7, n}\{i\}} \operatorname{cdet}_i\left((\mathbf{T}^*\mathbf{T})_{.i}\left(\mathbf{d}_{.j}^N\right)\right)_{\beta}^{\beta}}{\sum_{\beta \in J_{r_7, n}} |\mathbf{T}^*\mathbf{T}|_{\beta}^{\beta} \sum_{\alpha \in I_{r_6, r}} |\mathbf{NN}^*|_{\alpha}^{\alpha}},
\tag{18}
$$

or

$$
x_{ij}^{03} = \frac{\sum_{\alpha \in I_{r_6, r}\{j\}} \operatorname{rdet}_j\left((\mathbf{NN}^*)_{j.}\left(\mathbf{d}_{i.}^T\right)\right)_{\alpha}^{\alpha}}{\sum_{\beta \in J_{r_7, n}} |\mathbf{T}^*\mathbf{T}|_{\beta}^{\beta} \sum_{\alpha \in I_{r_6, r}} |\mathbf{NN}^*|_{\alpha}^{\alpha}},
\tag{19}
$$

where

$$
\mathbf{d}_{.j}^N = \left[\sum_{\alpha \in I_{r_6, r}\{f\}} \operatorname{rdet}_j\left((\mathbf{NN}^*)_{j.}\left(\widehat{c}_{q.}^{(2)}\right)\right)_{\alpha}^{\alpha}\right] \in \mathbb{H}^{n \times 1}, \quad q = 1, \dots, n,
$$

$$
\mathbf{d}_{i.}^T = \left[\sum_{\beta \in J_{r_7, n}\{i\}} \operatorname{cdet}_i\left((\mathbf{T}^*\mathbf{T})_{.i}\left(\widehat{c}_{.l}^{(2)}\right)\right)_{\beta}^{\beta}\right] \in \mathbb{H}^{1 \times r}, \quad l = 1, \dots, r,
$$

are the column vector and the row vector, respectively. $\widehat{c}_{q.}^{(2)}$ is the qth row and $\widehat{c}_{.l}^{(2)}$ is the lth column of $\widehat{\mathbf{C}}_2 = \mathbf{T}^*\mathbf{C}_2\mathbf{N}^*$. The following expression gives some simplify in computing. Since $\mathbf{T}^*\mathbf{T} = (\mathbf{R}_H\mathbf{A}_2)^* = \mathbf{A}_2^*\mathbf{R}_H^*\mathbf{R}_H\mathbf{A}_2 = \mathbf{A}_2^*\mathbf{R}_H\mathbf{A}_2$ and $\mathbf{R}_H = \mathbf{I} - \mathbf{HH}^\dagger = \mathbf{I} - \mathbf{A}_2\mathbf{L}_{A_1}(\mathbf{A}_2\mathbf{L}_{A_1})^\dagger = \mathbf{I} - \mathbf{A}_2(\mathbf{A}_2\mathbf{L}_{A_1})^\dagger$, then $\mathbf{T}^*\mathbf{T} = \mathbf{A}_2^*\left(\mathbf{I} - \mathbf{A}_2(\mathbf{A}_2\mathbf{L}_{A_1})^\dagger\right)\mathbf{A}_2$.

(iv) Using (3) for determinantal representations of $\mathbf{H}^\dagger$ and $\mathbf{T}^\dagger$ in the fourth term of (13), we obtain

$$
x_{ij}^{04} = \frac{\sum_{q=1}^{n}\sum_{z=1}^{n}\sum_{f=1}^{r}\sum_{\beta \in J_{r_5, n}\{i\}} \operatorname{cdet}_i\left((\mathbf{H}^*\mathbf{H})_{.i}\left(\mathbf{a}_{.q}^{(2,H)}\right)\right)_{\beta}^{\beta} \sum_{\beta \in J_{r_7, n}\{q\}} \operatorname{cdet}_q\left((\mathbf{T}^*\mathbf{T})_{.q}\left(\mathbf{a}_{.z}^{(2,T)}\right)\right)_{\beta}^{\beta} x_{zf}^{01} q_{fj}}{\sum_{\beta \in J_{r_5, n}} |\mathbf{H}^*\mathbf{H}|_{\beta}^{\beta} \sum_{\beta \in J_{r_7, n}} |\mathbf{T}^*\mathbf{T}|_{\beta}^{\beta}},
\tag{20}
$$

where $\mathbf{a}_{.i}^{(2,H)}$ and $\mathbf{a}_{.i}^{(2,T)}$ are the ith columns of the matrices $\mathbf{H}^*\mathbf{A}_2$ and $\mathbf{T}^*\mathbf{A}_2$, respectively; q_{fj} is the (f)th element of $\mathbf{Q}_{B_2}$ with the determinantal representation,

$$q_{fj} = \frac{\sum_{\alpha \in I_{r_4,r}\{j\}} \mathrm{rdet}_j \left(\left(\mathbf{B}_2\mathbf{B}_2^* \right)_{j.} \left(\ddot{\mathbf{b}}_{f.}^{(2)} \right) \right)_\alpha^\alpha}{\sum_{\alpha \in I_{r_4,r}} \left| \mathbf{B}_2\mathbf{B}_2^* \right|_\alpha^\alpha},$$

and $\ddot{\mathbf{b}}_{f.}^{(2)}$ is the fth row of $\mathbf{B}_2\mathbf{B}_2^*$. Note that $\mathbf{H}^*\mathbf{A}_2 = \mathbf{L}_{A_1}\mathbf{A}_2^*\mathbf{A}_2$ and $\mathbf{T}^*\mathbf{A}_2 = \mathbf{A}_2^*\mathbf{R}_H\mathbf{A}_2 = \mathbf{A}_2^*$ $\left(\mathbf{I} - \mathbf{A}_2(\mathbf{A}_2\mathbf{L}_{A_1})^+ \right)\mathbf{A}_2$.

(v) Similar to the previous case,

$$x_{ij}^{05} = \frac{\sum_{q=1}^n \sum_{f=1}^r \sum_{\beta \in J_{r_5,n}\{i\}} \mathrm{cdet}_i \left((\mathbf{H}^*\mathbf{H})_{.i} \left(\mathbf{a}_{.q}^{(2,H)} \right) \right)_\beta^\beta x_{qf}^{01} q_{fj}}{\sum_{\beta \in J_{r_5,n}} \left| \mathbf{H}^*\mathbf{H} \right|_\beta^\beta}, \tag{21}$$

(vi) Consider the sixth term by analogy to the fourth term. So,

$$x_{ij}^{06} = \frac{\sum_{q=1}^n \sum_{\beta \in J_{r_5,n}\{i\}} \mathrm{cdet}_i \left((\mathbf{H}^*\mathbf{H})_{.i} \left(\mathbf{a}_{.q}^{(2,H)} \right) \right)_\beta^\beta \varphi_{qj}}{\sum_{\beta \in J_{r_5,n}} \left| \mathbf{H}^*\mathbf{H} \right|_\beta^\beta \sum_{\beta \in J_{r_7,n}} \left| \mathbf{T}^*\mathbf{T} \right|_\beta^\beta \sum_{\alpha \in I_{r_4,r}} \left| \mathbf{B}_2\mathbf{B}_2^* \right|_\alpha^\alpha}, \tag{22}$$

where

$$\varphi_{qj} = \sum_{\beta \in J_{r_7,n}\{i\}} \mathrm{cdet}_q \left((\mathbf{T}^*\mathbf{T})_{.q} \left(\psi_{.j}^{B_2} \right) \right)_\beta^\beta, \tag{23}$$

or

$$\varphi_{qj} = \sum_{\alpha \in I_{r_4,r}\{j\}} \mathrm{rdet}_j \left(\left(\mathbf{B}_2\mathbf{B}_2^* \right)_{j.} \left(\psi_{q.}^T \right) \right)_\alpha^\alpha, \tag{24}$$

and

$$\psi_{.j}^{B_2} = \left[\sum_{\alpha \in I_{r_1,r}\{f\}} \mathrm{rdet}_j \left(\left(\mathbf{B}_2\mathbf{B}_2^* \right)_{j.} \left(\check{\mathbf{c}}_{q.}^{(2)} \right) \right)_\alpha^\alpha \right] \in \mathbb{H}^{1 \times n}, \quad q = 1, \ldots, n,$$

$$\psi_{q.}^T = \left[\sum_{\beta \in J_{r_7,n}\{q\}} \mathrm{cdet}_q \left((\mathbf{T}^*\mathbf{T})_{.q} \left(\check{\mathbf{c}}_{.l}^{(2)} \right) \right)_\beta^\beta \right] \in \mathbb{H}^{r \times 1}, \quad l = 1, \ldots, r,$$

are the column vector and the row vector, respectively. $\check{c}_{q.}(2)$ and $\check{c}_{.l}(2)$ are the qth row and the lth column of $\check{C}_2 = T^*C_2B_2^*$ for all $i = 1, ..., n$ and $j = 1, ..., p$.

(vii) Using (3) for determinantal representations of and $T^\dagger$ and (2) for $N^\dagger$ in the seventh term of (13), we obtain

$$
x_{ij}^{07} = \frac{\sum_{q=1}^{n} \sum_{f=1}^{r} \sum_{\beta \in J_{r_7,n}\{i\}} \mathrm{cdet}_i\left((T^*T)_{.i}\left(a_{.q}^{(2,T)}\right)\right)_\beta^\beta x_{qf}^{01} \sum_{\alpha \in I_{r_6,r}\{j\}} \mathrm{rdet}_j\left((NN^*)_{j.}\left(b_{f.}^{(2,N)}\right)\right)_\alpha^\alpha}{\sum_{\beta \in J_{r_7,n}} |T^*T|_\beta^\beta \sum_{\alpha \in I_{r_6,r}} |NN^*|_\alpha^\alpha},
$$

$$(25)$$

where $a_{.q}^{(2,T)}$ and $b_{f.}^{(2,N)}$ are the qth column of T^*A_2 and the fth row of $B_2N^* = B_2B_2^*R_{B_1}$, respectively.

Hence, we prove the following theorem.

Theorem 3.1. Let $A_1 \in \mathbb{H}_{r_1}^{m \times n}$, $B_1 \in \mathbb{H}_{r_2}^{r \times s}$, $A_2 \in \mathbb{H}_{r_3}^{k \times n}$, $B_2 \in \mathbb{H}_{r_4}^{r \times p}$, $\mathrm{rank}H = \mathrm{rank}(A_2L_{A_1}) = r_5$, $\mathrm{rank}N = (R_{B_1}B_2) = r_6$, and $\mathrm{rank}T = (R_HA_2) = r_7$. Then, for the partial solution (13), $X_0 = \left(x_{ij}^0\right) \in \mathbb{H}^{n \times r}$, of the system (1), we have,

$$
x_{ij}^0 = \sum_\delta x_{ij}^{0\delta}, \tag{26}
$$

where the term x_{ij}^{01} has the determinantal representations (14) and (15), $x_{ij}^{02}-(16)$ and (17), $x_{ij}^{03}-(18)$ and (19), $x_{ij}^{04}-(20)$, $x_{ij}^{05}-(21)$, $x_{ij}^{06}-(23)$ and (24), and $x_{ij}^{07}-(25)$.

4. Cramer's rules for special cases of (1)

In this section, we consider special cases of (1) when one or two equations are one-sided. Let in Eq.(1), the matrix B_1 is vanished. Then, we have the system

$$
\begin{cases} A_1X = C_1, \\ A_2XB_2 = C_2. \end{cases} \tag{27}
$$

The following lemma is extended to matrices with quaternion entries.

Lemma 4.1. [7] Let $A_1 \in \mathbb{H}^{m \times n}$, $C_1 \in \mathbb{H}^{m \times r}$, $A_2 \in \mathbb{H}^{k \times n}$, $B_2 \in \mathbb{H}^{r \times p}$, and $C_2 \in \mathbb{H}^{k \times p}$ be given and $X \in \mathbb{H}^{n \times r}$ is to be determined. Put $H = A_2L_{A_1}$. Then, the following statements are equivalent:

i.   *System (27) is consistent.*

ii. $\mathbf{R}_{A_1}\mathbf{C}_1 = 0,\ \mathbf{R}_H\big(\mathbf{C}_2 - \mathbf{A}_2\mathbf{A}_1^{\dagger}\mathbf{C}_1\mathbf{B}_2\big) = 0,\ \mathbf{C}_2\mathbf{L}_{B_2} = 0.$

iii. $\operatorname{rank}\begin{bmatrix}\mathbf{A}_1 & \mathbf{C}_1\end{bmatrix} = \operatorname{rank}\begin{bmatrix}\mathbf{A}_1\end{bmatrix},\ \operatorname{rank}\begin{bmatrix}\mathbf{C}_2 \\ \mathbf{B}_2\end{bmatrix} = \operatorname{rank}\begin{bmatrix}\mathbf{B}_2\end{bmatrix},\ \operatorname{rank}\begin{bmatrix}\mathbf{A}_1 & \mathbf{C}_1\mathbf{B}_2 \\ \mathbf{A}_2 & \mathbf{C}_2\end{bmatrix} = \operatorname{rank}\begin{bmatrix}\mathbf{A}_1 \\ \mathbf{A}_2\end{bmatrix}.$

In this case, the general solution of (27) can be expressed as

$$\mathbf{X} = \mathbf{A}_1^{\dagger}\mathbf{C}_1 + \mathbf{L}_{A_1}\mathbf{H}^{\dagger}\big(\mathbf{C}_2 - \mathbf{A}_2\mathbf{A}_1^{\dagger}\mathbf{C}_1\mathbf{B}_2\big)\mathbf{B}_2^{\dagger} + \mathbf{L}_{A_1}\mathbf{L}_H\mathbf{Z}_1 + \mathbf{L}_{A_1}\mathbf{W}_1\mathbf{R}_{B_2}, \tag{28}$$

where $\mathbf{Z}_1$ and $\mathbf{W}_1$ are the arbitrary matrices over $\mathbb{H}$ with appropriate sizes.

Since by (9), $\mathbf{L}_{A_1}\mathbf{H}^{\dagger} = \mathbf{L}_{A_1}\big(\mathbf{A}_2\mathbf{L}_{A_1}\big)^{\dagger} = \big(\mathbf{A}_2\mathbf{L}_{A_1}\big)^{\dagger} = \mathbf{H}^{\dagger}$, then we have some simplification of (28),

$$\mathbf{X} = \mathbf{A}_1^{\dagger}\mathbf{C}_1 + \mathbf{H}^{\dagger}\mathbf{C}_2\mathbf{B}_2^{\dagger} - \mathbf{H}^{\dagger}\mathbf{A}_2\mathbf{A}_1^{\dagger}\mathbf{C}_1\mathbf{B}_2\mathbf{B}_2^{\dagger} + \mathbf{L}_{A_1}\mathbf{L}_H\mathbf{Z}_1 + \mathbf{L}_{A_1}\mathbf{W}_1\mathbf{R}_{B_2}.$$

By putting $\mathbf{Z}_1=\mathbf{W}_1=0$, there is the following partial solution of (27),

$$\mathbf{X}_0 = \mathbf{A}_1^{\dagger}\mathbf{C}_1 + \mathbf{H}^{\dagger}\mathbf{C}_2\mathbf{B}_2^{\dagger} - \mathbf{H}^{\dagger}\mathbf{A}_2\mathbf{A}_1^{\dagger}\mathbf{C}_1\mathbf{B}_2\mathbf{B}_2^{\dagger}. \tag{29}$$

Theorem   4.1. *Let* $\mathbf{A}_1 = \big(a_{ij}^{(1)}\big) \in \mathbb{H}_{r_1}^{m\times n},\quad \mathbf{A}_2 = \big(a_{ij}^{(2)}\big) \in \mathbb{H}_{r_2}^{k\times n},\quad \mathbf{B}_2 = \big(b_{ij}^{(2)}\big) \in \mathbb{H}_{r_3}^{r\times p},$
$\mathbf{C}_1 = \big(c_{ij}^{(1)}\big) \in \mathbb{H}^{m\times r},\quad and \quad \mathbf{C}_2 = \big(c_{ij}^{(2)}\big) \in \mathbb{H}^{k\times p},\quad and \quad there \quad exist \quad \mathbf{A}_1^{\dagger} = \big(a_{ij}^{(1),\,\dagger}\big) \in \mathbb{H}^{n\times m},$
$\mathbf{B}_2^{\dagger} = \big(b_{ij}^{(2),\,\dagger}\big) \in \mathbb{H}^{p\times r},\ and\ \mathbf{H}^{\dagger} = \big(h_{ij}^{\dagger}\big) \in \mathbb{H}^{n\times k}.\ Let\ \operatorname{rank}\mathbf{H} = \min\{\operatorname{rank}\mathbf{A}_2,\operatorname{rank}\mathbf{L}_{A_1}\} = r_4.\ Denote$
$\mathbf{A}_1^{\dagger}\mathbf{C}_1=:\widehat{\mathbf{C}}_1 = \big(\widehat{c}_{ij}^{(1)}\big) \in \mathbb{H}^{n\times r},\ \mathbf{H}^{*}\mathbf{C}_2\mathbf{B}_2^{*}=:\widehat{\mathbf{C}}_2 = \big(\widehat{c}_{ij}^{(2)}\big) \in \mathbb{H}^{n\times r},\ \mathbf{H}^{*}\mathbf{A}_2\mathbf{A}_1^{*}=:\widehat{\mathbf{A}}_2 = \big(\widehat{a}_{ij}^{(2)}\big) \in \mathbb{H}^{n\times m},\ and$
$\mathbf{C}_1\mathbf{Q}_{B_2}=:\widehat{\mathbf{Q}} = \big(\widehat{q}_{ij}\big) \in \mathbb{H}^{m\times p}.\ Then,\ the\ partial\ solution\ (29),\ \mathbf{X}_0 = \big(x_{ij}^0\big) \in \mathbb{H}^{n\times r},\ possesses\ the\ following$
determinantal representations,

$$x_{ij}^0 = \frac{\sum_{\beta \in J_{r_1,n}\{i\}}\operatorname{cdet}_i\left(\big(\mathbf{A}_1^{*}\mathbf{A}_1\big)_{.i}\big(\widehat{\mathbf{c}}_{.j}^{(1)}\big)\right)_{\beta}^{\beta}}{\sum_{\beta \in J_{r_1,n}}\big|\mathbf{A}_1^{*}\mathbf{A}_1\big|_{\beta}^{\beta}} + \frac{d_{ij}^{(\lambda)}}{\sum_{\beta \in J_{r_4,n}}\big|\mathbf{H}^{*}\mathbf{H}\big|_{\beta}^{\beta}\sum_{\alpha \in I_{r_3,r}}\big|\mathbf{B}_2\mathbf{B}_2^{*}\big|_{\alpha}^{\alpha}}$$
$$- \frac{\sum_{l=1}^{m}g_{il}^{(\mu)}\sum_{\alpha \in I_{r_3,r}\{j\}}\operatorname{rdet}_j\left(\big(\mathbf{B}_2\mathbf{B}_2^{*}\big)_{j.}\big(\widehat{\mathbf{q}}_{l.}\big)\right)_{\alpha}^{\alpha}}{\sum_{\beta \in J_{r_4,n}}\big|\mathbf{H}^{*}\mathbf{H}\big|_{\beta}^{\beta}\sum_{\alpha \in I_{r_1,m}}\big|\mathbf{A}_1\mathbf{A}_1^{*}\big|_{\alpha}^{\alpha}\sum_{\alpha \subset I_{r_3,r}}\big|\mathbf{B}_2\mathbf{B}_2^{*}\big|_{\alpha}^{\alpha}}$$

for all $\lambda = 1, 2$ *and* $\mu = 1, 2.$ *Here*

$$d_{ij}^{(1)} := \sum_{\alpha \in I_{r_3,r}\{j\}}\operatorname{rdet}_j\left(\big(\mathbf{D}_{\lambda}\mathbf{B}_2^{*}\big)_{j.}\big(\mathbf{v}_{i.}^{(1)}\big)\right)_{\alpha}^{\alpha},\qquad g_{il}^{(1)} := \sum_{\alpha \in I_{r_1,m}\{l\}}\operatorname{rdet}_l\left(\big(\mathbf{A}_1\mathbf{A}_1^{*}\big)_{l.}\big(\mathbf{u}_{i.}^{(1)}\big)\right)_{\alpha}^{\alpha},$$

and the row-vectors $\mathbf{v}_{i.}^{(1)} = \big(v_{i1}^{(1)}, \ldots, v_{ir}^{(1)}\big)$ *and* $\mathbf{u}_{i.}^{(1)} = \big(u_{i1}^{(1)}, \ldots, u_{im}^{(1)}\big)$ *such that*

$$v_{it}^{(1)} := \sum_{\beta \in J_{r_4,n}\{i\}} \mathrm{cdet}_i \left((\mathbf{H}^*\mathbf{H})_{.i} \left(\widehat{\mathbf{c}}_{.t}^{(2)} \right) \right)_{\beta}^{\beta}, \quad u_{iz}^{(1)} := \sum_{\beta \in J_{r_4,n}\{i\}} \mathrm{cdet}_i \left((\mathbf{H}^*\mathbf{H})_{.i} \left(\widehat{\mathbf{a}}_{.z}^{(2)} \right) \right)_{\beta}^{\beta}.$$

In another case,

$$d_{ij}^{(2)} := \sum_{\beta \in J_{r_4,n}\{i\}} \mathrm{cdet}_i \left((\mathbf{H}^*\mathbf{H})_{.i} \left(\mathbf{v}_{.j}^{(2)} \right) \right)_{\beta}^{\beta}, \quad g_{il}^{(2)} := \sum_{\beta \in J_{r_4,n}\{i\}} \mathrm{cdet}_i \left((\mathbf{H}^*\mathbf{H})_{.i} \left(\mathbf{u}_{.l}^{(2)} \right) \right)_{\beta}^{\beta}.$$

and the column-vectors $\mathbf{v}_{.j}^{(2)} = \left(v_{1j}^{(2)}, \ldots, v_{nj}^{(2)} \right)$ and $\mathbf{u}_{.l}^{(2)} = \left(u_{1l}^{(2)}, \ldots, u_{nl}^{(2)} \right)$ such that

$$v_{qj}^{(2)} := \sum_{\alpha \in I_{r_3,r}\{j\}} \mathrm{rdet}_j \left((\mathbf{B}_2\mathbf{B}_2^*)_{j.} \left(\widehat{\mathbf{c}}_{q.}^{(2)} \right) \right)_{\alpha}^{\alpha}, \quad u_{ql}^{(2)} := \sum_{\alpha \in I_{r_1,m}\{l\}} \mathrm{rdet}_l \left((\mathbf{A}_1\mathbf{A}_1^*)_{l.} \left(\widehat{\mathbf{a}}_{q.}^{(2)} \right) \right)_{\alpha}^{\alpha}.$$

Proof. The proof is similar to the proof of Theorem 3.1.

Let in Eq.(1), the matrix $\mathbf{A}_1$ is vanished. Then, we have the system,

$$\begin{cases} \mathbf{X}\mathbf{B}_1 = \mathbf{C}_1, \\ \mathbf{A}_2\mathbf{X}\mathbf{B}_2 = \mathbf{C}_2. \end{cases} \tag{30}$$

The following lemma is extended to matrices with quaternion entries as well.

Lemma 4.2. [7] *Let* $\mathbf{B}_1 \in \mathbb{H}^{r \times s}$, $\mathbf{C}_1 \in \mathbb{H}^{n \times s}$, $\mathbf{A}_2 \in \mathbb{H}^{k \times n}$, $\mathbf{B}_2 \in \mathbb{H}^{r \times p}$, *and* $\mathbf{C}_2 \in \mathbb{H}^{k \times p}$ *be given and* $\mathbf{X} \in \mathbb{H}^{n \times r}$ *is to be determined. Put* $\mathbf{N} = \mathbf{R}_{B_1}\mathbf{B}_2$. *Then, the following statements are equivalent:*

i.   *System (30) is consistent.*

ii. $\mathbf{R}_{A_2}\mathbf{C}_2 = 0,\ (\mathbf{C}_2 - \mathbf{A}_2\mathbf{C}_1\mathbf{B}_1^{\dagger}\mathbf{B}_2)\mathbf{L}_N = 0,\ \mathbf{C}_2\mathbf{L}_{B_2} = 0.$

iii. $\mathrm{rank}[\,\mathbf{A}_2 \quad \mathbf{C}_2\,] = \mathrm{rank}[\mathbf{A}_2],\ \mathrm{rank}\begin{bmatrix} \mathbf{C}_1 \\ \mathbf{B}_1 \end{bmatrix} = \mathrm{rank}[\mathbf{B}_1],\ \mathrm{rank}\begin{bmatrix} \mathbf{C}_2 & \mathbf{A}_2\mathbf{C}_1 \\ \mathbf{B}_2 & \mathbf{B}_1 \end{bmatrix} = \mathrm{rank}[\,\mathbf{B}_2 \quad \mathbf{B}_1\,].$

In this case, the general solution of (30) can be expressed as

$$\mathbf{X} = \mathbf{C}_1\mathbf{B}_1^{\dagger} + \mathbf{A}_2^{\dagger}(\mathbf{C}_2 - \mathbf{A}_2\mathbf{C}_1\mathbf{B}_1^{\dagger}\mathbf{B}_2)\mathbf{N}^{\dagger}\mathbf{R}_{B_1} + \mathbf{L}_{A_2}\mathbf{W}_2\mathbf{R}_{B_1} + \mathbf{Z}_2\mathbf{R}_N\mathbf{R}_{B_1}, \tag{31}$$

where $\mathbf{Z}_2$ and $\mathbf{W}_2$ are the arbitrary matrices over $\mathbb{H}$ with appropriate sizes.

Since by (10), $\mathbf{N}^{\dagger}\mathbf{R}_{B_1} = (\mathbf{R}_{B_1}\mathbf{B}_2)^{\dagger}\mathbf{R}_{B_1} = \mathbf{N}^{\dagger}$, then some simplification of (31) can be derived,

$$\mathbf{X} = \mathbf{C}_1\mathbf{B}_1^{\dagger} + \mathbf{A}_2^{\dagger}\mathbf{C}_2\mathbf{N}^{\dagger} - \mathbf{A}_2\mathbf{C}_1\mathbf{B}_1^{\dagger}\mathbf{B}_2\mathbf{N}^{\dagger} + \mathbf{L}_{A_2}\mathbf{W}_2\mathbf{R}_{B_1} + \mathbf{Z}_2\mathbf{R}_N\mathbf{R}_{B_1}.$$

By putting $\mathbf{Z}_2=\mathbf{W}_2=0$, there is the following partial solution of (30),

$$\mathbf{X}_0 = \mathbf{C}_1\mathbf{B}_1^{\dagger} + \mathbf{A}_2^{\dagger}\mathbf{C}_2\mathbf{N}^{\dagger} - \mathbf{A}_2^{\dagger}\mathbf{A}_2\mathbf{C}_1\mathbf{B}_1^{\dagger}\mathbf{B}_2\mathbf{N}^{\dagger}. \tag{32}$$

The following theorem on determinantal representations of (29) can be proven similar to the proof of Theorem 3.1 as well.

Theorem   4.2. Let $\mathbf{B}_1 = \left(b_{ij}^{(1)}\right) \in \mathbb{H}_{r_1}^{r \times s}$, $\mathbf{A}_2 = \left(a_{ij}^{(2)}\right) \in \mathbb{H}_{r_2}^{k \times n}$, $\mathbf{B}_2 = \left(b_{ij}^{(2)}\right) \in \mathbb{H}_{r_3}^{r \times p}$, $\mathbf{C}_1 = \left(c_{ij}^{(1)}\right) \in \mathbb{H}^{n \times s}$, and $\mathbf{C}_2 = \left(c_{ij}^{(2)}\right) \in \mathbb{H}^{k \times p}$, and there exist $\mathbf{B}_1^\dagger = \left(b_{ij}^{(1),+}\right) \in \mathbb{H}^{s \times r}$, $\mathbf{A}_2^\dagger = \left(a_{ij}^{(2),+}\right) \in \mathbb{H}^{n \times k}$, $\mathbf{N}^\dagger = \left(n_{ij}^\dagger\right) \in \mathbb{H}^{p \times r}$. Let $\operatorname{rank}\mathbf{N} = \min\{\operatorname{rank}\mathbf{B}_2, \operatorname{rank}\mathbf{R}_{\mathbf{B}_1}\} = r_4$. Denote $\mathbf{C}_1\mathbf{B}_1^* =: \tilde{\mathbf{C}}_1 = \left(\tilde{c}_{ij}^{(1)}\right) \in \mathbb{H}^{n \times r}$,

$\mathbf{A}_2^*\mathbf{C}_2\mathbf{N}^* =: \tilde{\mathbf{C}}_2 = \left(\tilde{c}_{ij}^{(2)}\right) \in \mathbb{H}^{n \times r}$, $\mathbf{B}_1^*\mathbf{B}_2\mathbf{N}^* =: \tilde{\mathbf{B}}_2 = \left(\tilde{b}_{ij}^{(2)}\right) \in \mathbb{H}^{s \times r}$, and $\mathbf{P}_{\mathbf{A}_2}\mathbf{C}_1 =: \tilde{\mathbf{P}} = \left(\tilde{p}_{ij}\right) \in \mathbb{H}^{n \times s}$. Then,

the partial solution (32), $\mathbf{X}_0 = \left(x_{ij}^0\right) \in \mathbb{H}^{n \times r}$, possesses the following determinantal representations,

$$x_{ij}^0 = \frac{\sum_{\alpha \in I_{r_1, r}\{j\}} \operatorname{rdet}_j\left(\left(\mathbf{B}_1\mathbf{B}_1^*\right)_{j.}\left(\tilde{\mathbf{c}}_{i.}^{(1)}\right)\right)_\alpha}{\sum_{\alpha \in I_{r_1, r}} |\mathbf{B}_1\mathbf{B}_1^*|_\alpha} + \frac{d_{ij}^{(\lambda)}}{\sum_{\beta \in I_{r_2, n}} |\mathbf{A}_2^*\mathbf{A}_2|_\beta \sum_{\alpha \in I_{r_4, r}} |\mathbf{N}\mathbf{N}^*|_\alpha}$$

$$- \frac{\sum_{z=1}^s \sum_{\beta \in I_{r_2, n}\{i\}} \operatorname{cdet}_i\left(\left(\mathbf{A}_2^*\mathbf{A}_2\right)_{.i}\left(\tilde{\mathbf{p}}_{.z}\right)\right)_\beta g_{zj}^{(\mu)}}{\sum_{\beta \in I_{r_2, n}} |\mathbf{A}_2^*\mathbf{A}_2|_\beta \sum_{\beta \in I_{r_1, s}} |\mathbf{B}_1^*\mathbf{B}_1|_\beta \sum_{\alpha \in I_{r_4, r}} |\mathbf{N}\mathbf{N}^*|_\alpha}$$

for all $\lambda = 1, 2$ and $\mu = 1, 2$. Here

$$d_{ij}^{(1)} := \sum_{\alpha \in I_{r_3, r}\{j\}} \operatorname{rdet}_j\left(\left(\mathbf{N}\mathbf{N}^*\right)_{j.}\left(\varphi_{i.}^{(1)}\right)\right)_\alpha, \quad g_{il}^{(1)} := \sum_{\alpha \in I_{r_4, r}\{j\}} \operatorname{rdet}_j\left(\left(\mathbf{N}\mathbf{N}^*\right)_{j.}\left(\psi_{z.}\right)\right)_\alpha,$$

and the row-vectors $\varphi_{i.}^{(1)} = \left(\varphi_{i1}^{(1)}, \ldots, \varphi_{ir}^{(1)}\right)$ and $\psi_{i.}^{(1)} = \left(\psi_{z1}^{(1)}, \ldots, \psi_{zr}^{(1)}\right)$ such that

$$\varphi_{it}^{(1)} = \sum_{\beta \in I_{r_2, n}\{i\}} \operatorname{cdet}_i\left(\left(\mathbf{A}_2^*\mathbf{A}_2\right)_{.i}\left(c_{.t}^{(2)}\right)\right)_\beta, \quad \psi_{zv}^{(1)} = \sum_{\beta \in I_{r_1, n}\{z\}} \operatorname{cdet}_z\left(\left(\mathbf{B}_1^*\mathbf{B}_1\right)_{.i}\left(b_{.v}^{(2)}\right)\right)_\beta.$$

In another case,

$$d_{ij}^{(2)} := \sum_{\beta \in I_{r_2, n}\{i\}} \operatorname{cdet}_i\left(\left(\mathbf{A}_2^*\mathbf{A}_2\right)_{.i}\left(\varphi_{.j}^{(2)}\right)\right)_\beta, \quad g_{zj}^{(2)} := \sum_{\beta \in I_{r_1, n}\{z\}} \operatorname{cdet}_z\left(\left(\mathbf{B}_1^*\mathbf{B}_1\right)_{.z}\left(\psi_{.j}^{(2)}\right)\right)_\beta,$$

and the column-vectors $\varphi_{.j}^{(2)} = \left(\varphi_{1j}^{(2)}, \ldots, \varphi_{nj}^{(2)}\right)$ and $\psi_{.j}^{(2)} = \left(\psi_{1j}^{(2)}, \ldots, \psi_{sj}^{(2)}\right)$ such that

$$\varphi_{qj}^{(2)} = \sum_{\alpha \in I_{r_4, r}\{j\}} \operatorname{rdet}_j\left(\left(\mathbf{N}\mathbf{N}^*\right)_{j.}\left(c_{q.}^{(2)}\right)\right)_\alpha, \quad \psi_{uj}^{(2)} := \sum_{\alpha \in I_{r_4, r}\{j\}} \operatorname{rdet}_j\left(\left(\mathbf{N}\mathbf{N}^*\right)_{j.}\left(b_{u.}^{(2)}\right)\right)_\alpha.$$

Now, suppose that the both equations of (1) are one-sided. Let in Eq.(1), the matrices $\mathbf{B}_1$ and $\mathbf{A}_2$ are vanished. Then, we have the system

$$\begin{cases} \mathbf{A}_1\mathbf{X} = \mathbf{C}_1, \\ \mathbf{X}\mathbf{B}_2 = \mathbf{C}_2. \end{cases} \tag{33}$$

The following lemma is extended to matrices with quaternion entries.

Lemma 4.3. [31] *Let* $\mathbf{A}_1 \in \mathbb{H}^{m \times n}$, $\mathbf{B}_2 \in \mathbb{H}^{r \times p}$, $\mathbf{C}_1 \in \mathbb{H}^{m \times r}$, *and* $\mathbf{C}_2 \in \mathbb{H}^{n \times p}$ *be given and* $\mathbf{X} \in \mathbb{H}^{n \times r}$ *is to be determined. Then, the system (33) is consistent if and only if* $\mathbf{R}_{A_1}\mathbf{C}_1 = 0$, $\mathbf{C}_2\mathbf{L}_{B_2} = 0$, *and* $\mathbf{A}_1\mathbf{C}_2 = \mathbf{C}_1\mathbf{B}_2$. *Under these conditions, the general solution to (33) can be established as*

$$\mathbf{X} = \mathbf{A}_1^\dagger \mathbf{C}_1 + \mathbf{L}_{A_1}\mathbf{C}_2\mathbf{B}_2^\dagger + \mathbf{L}_{A_1}\mathbf{U}\mathbf{R}_{B_2}, \tag{34}$$

where $\mathbf{U}$ is a free matrix over $\mathbb{H}$ with a suitable shape.

Due to the consistence conditions, Eq. (34) can be expressed as follows:

$$\begin{aligned} \mathbf{X} &= \mathbf{C}_2\mathbf{B}_2^\dagger + \mathbf{A}_1^\dagger(\mathbf{C}_1 - \mathbf{A}_1\mathbf{C}_2\mathbf{B}_2^\dagger) + \mathbf{L}_{A_1}\mathbf{U}\mathbf{R}_{B_2} \\ &= \mathbf{C}_2\mathbf{B}_2^\dagger + \mathbf{A}_1^\dagger(\mathbf{C}_1 - \mathbf{C}_1\mathbf{B}_2\mathbf{B}_2^\dagger) + \mathbf{L}_{A_1}\mathbf{U}\mathbf{R}_{B_2} = \mathbf{C}_2\mathbf{B}_2^\dagger + \mathbf{A}_1^\dagger\mathbf{C}_1\mathbf{R}_{B_2} + \mathbf{L}_{A_1}\mathbf{U}\mathbf{R}_{B_2}, \end{aligned}$$

Consequently, the partial solution $\mathbf{X}^0$ to (33) is given by

$$\mathbf{X}^0 = \mathbf{A}_1^\dagger\mathbf{C}_1 + \mathbf{L}_{A_1}\mathbf{C}_2\mathbf{B}_2^\dagger, \tag{35}$$

or

$$\mathbf{X}^0 = \mathbf{C}_2\mathbf{B}_2^\dagger + \mathbf{A}_1^\dagger\mathbf{C}_1\mathbf{R}_{B_2}. \tag{36}$$

Due to the expression (35), the following theorem can be proven similar to the proof of Theorem 3.1.

Theorem 4.3. *Let* $\mathbf{A}_1 = \left(a_{ij}^{(1)}\right) \in \mathbb{H}_{r_1}^{m \times n}$, $\mathbf{B}_2 = \left(b_{ij}^{(2)}\right) \in \mathbb{H}_{r_2}^{r \times p}$, $\mathbf{C}_1 = \left(c_{ij}^{(1)}\right) \in \mathbb{H}^{m \times r}$, *and* $\mathbf{C}_2 = \left(c_{ij}^{(2)}\right) \in \mathbb{H}^{n \times r}$, *and there exist* $\mathbf{A}_1^\dagger = \left(a_{ij}^{(1),\dagger}\right) \in \mathbb{H}^{n \times m}$, $\mathbf{B}_2^\dagger = \left(b_{ij}^{(2),\dagger}\right) \in \mathbb{H}^{p \times r}$, *and* $\mathbf{L}_{A_1} = \mathbf{I} - \mathbf{A}_1^\dagger\mathbf{A}_1 =: \left(l_{ij}\right) \in \mathbb{H}^{n \times n}$. *Denote* $\mathbf{A}_1^*\mathbf{C}_1 =: \widehat{\mathbf{C}}_1 = \left(\hat{c}_{ij}^{(1)}\right) \in \mathbb{H}^{n \times r}$ *and* $\mathbf{L}_{A_1}\mathbf{C}_2\mathbf{B}_2^* =: \widehat{\mathbf{C}}_2 = \left(\hat{c}_{ij}^{(2)}\right) \in \mathbb{H}^{n \times r}$. *Then, the partial solution (35),* $\mathbf{X}^0 = \left(x_{ij}^0\right) \in \mathbb{H}^{n \times s}$, *possesses the following determinantal representation,*

$$x_{ij}^0 = \frac{\sum_{\beta \in J_{r_1,n}\{i\}} \mathrm{cdet}_i\left(\left(\mathbf{A}_1^*\mathbf{A}_1\right)_{.i}\left(\hat{c}_{.j}^{(1)}\right)\right)_\beta^\beta}{\sum_{\beta \in J_{r_1,n}} \left|\mathbf{A}_1^*\mathbf{A}_1\right|_\beta^\beta} + \frac{\sum_{\alpha \in I_{r_2,r}\{j\}} \mathrm{rdet}_j\left(\left(\mathbf{B}_2\mathbf{B}_2^*\right)_{j.}\left(\hat{c}_{i.}^{(2)}\right)\right)_\alpha^\alpha}{\sum_{\alpha \in I_{r_2,r}} \left|\mathbf{B}_2\mathbf{B}_2^*\right|_\alpha^\alpha}, \tag{37}$$

where $\hat{c}_{.j}^{(1)}$ is the jth column of $\widehat{\mathbf{C}}_1$ and $\hat{c}_{i.}^{(2)}$ is the ith row of $\widehat{\mathbf{C}}_2$.

Remark 4.1. *In accordance to the expression (36), we obtain the same representations, but with the denotations,* $\mathbf{C}_2\mathbf{B}_2^* =: \widehat{\mathbf{C}}_2 = \left(\hat{c}_{ij}^{(2)}\right) \in \mathbb{H}^{n \times r}$ *and* $\mathbf{A}_1^*\mathbf{C}_1\mathbf{R}_{B_2} =: \widehat{\mathbf{C}}_1 = \left(\hat{c}_{ij}^{(2)}\right) \in \mathbb{H}^{n \times r}$.

Let in Eq.(1), the matrices $\mathbf{B}_1$ and $\mathbf{B}_2$ are vanished. Then, we have the system

$$\begin{cases} \mathbf{A}_1\mathbf{X} = \mathbf{C}_1, \\ \mathbf{A}_2\mathbf{X} = \mathbf{C}_2. \end{cases} \tag{38}$$

Lemma 4.4. [7] *Suppose that* $\mathbf{A}_1 \in \mathbb{H}^{m\times n}$, $\mathbf{C}_1 \in \mathbb{H}^{m\times r}$, $\mathbf{A}_2 \in \mathbb{H}^{k\times n}$, *and* $\mathbf{C}_2 \in \mathbb{H}^{k\times r}$ *are known and* $\mathbf{X} \in \mathbb{H}^{n\times r}$ *is unknown,* $\mathbf{H} = \mathbf{A}_2\mathbf{L}_{A_1}$, $\mathbf{T} = \mathbf{R}_H\mathbf{A}_2$. *Then, the system (38) is consistent if and only if* $\mathbf{A}_i\mathbf{A}_i^\dagger\mathbf{C}_i = \mathbf{C}_i$, *for all* $i = 1, 2$ *and* $\mathbf{T}\left(\mathbf{A}_2^\dagger\mathbf{C}_2 - \mathbf{A}_1^\dagger\mathbf{C}_1\right) = 0$. *Under these conditions, the general solution to (38) can be established as*

$$\mathbf{X} = \mathbf{A}_1^\dagger\mathbf{C}_1 + \mathbf{L}_{A_1}\mathbf{H}^\dagger\mathbf{A}_2\left(\mathbf{A}_2^\dagger\mathbf{C}_2 - \mathbf{A}_1^\dagger\mathbf{C}_1\right) + \mathbf{L}_{A_1}\mathbf{L}_H\mathbf{Y}, \tag{39}$$

where $\mathbf{Y}$ is an arbitrary matrix over $\mathbb{H}$ with an appropriate size.

Using (10) and the consistency conditions, we simplify (39) accordingly, $\mathbf{X}^0 = \mathbf{A}_1^\dagger\mathbf{C}_1 + \mathbf{H}^\dagger\mathbf{C}_2 - \mathbf{H}^\dagger\mathbf{A}_2\mathbf{A}_1^\dagger\mathbf{C}_1 + \mathbf{L}_{A_1}\mathbf{L}_H\mathbf{Y}$. Consequently, the following partial solution of (39) will be considered

$$\mathbf{X}^0 = \mathbf{A}_1^\dagger\mathbf{C}_1 + \mathbf{H}^\dagger\mathbf{C}_2 - \mathbf{H}^\dagger\mathbf{A}_2\mathbf{A}_1^\dagger\mathbf{C}_1. \tag{40}$$

In the following theorem, we give the determinantal representations of (40).

Theorem 4.4. *Let* $\mathbf{A}_1 = \left(a_{ij}^{(1)}\right) \in \mathbb{H}^{m\times n}$, $\mathbf{A}_2 = \left(a_{ij}^{(2)}\right) \in \mathbb{H}_{r_1}^{k\times n}$, $\mathbf{C}_1 = \left(c_{ij}^{(1)}\right) \in \mathbb{H}^{m\times r}$, $\mathbf{C}_2 = \left(c_{ij}^{(2)}\right) \in \mathbb{H}^{k\times r}$, *and there exist* $\mathbf{A}_1^\dagger = \left(a_{ij}^{(1),\,+}\right) \in \mathbb{H}^{n\times m}$, $\mathbf{H}_2^\dagger = \left(h_{ij}^\dagger\right) \in \mathbb{H}^{n\times s}$. *Let* $\mathrm{rank}\,\mathbf{H} = \min\{\mathrm{rank}\,\mathbf{A}_2, \mathrm{rank}\,\mathbf{L}_{A_1}\} = r_3$. *Denote* $\mathbf{A}_1^*\mathbf{C}_1 =: \widehat{\mathbf{C}}_1 = \left(\widehat{c}_{ij}^{(1)}\right) \in \mathbb{H}^{n\times r}$, $\mathbf{H}^*\mathbf{C}_2 =: \widehat{\mathbf{C}}_2 = \left(\widehat{c}_{ij}^{(2)}\right) \in \mathbb{H}^{n\times r}$, *and* $\mathbf{H}^*\mathbf{A}_2 =: \widehat{\mathbf{A}}_2 = \left(\widehat{a}_{ij}^{(2)}\right) \in \mathbb{H}^{n\times n}$. *Then,* $\mathbf{X}^0 = \left(x_{ij}^0\right) \in \mathbb{H}^{n\times r}$ *possesses the following determinantal representation,*

$$x_{ij}^0 = \frac{\sum_{\beta\in J_{r_1,n}\{i\}}\mathrm{cdet}_i\left(\left(\mathbf{A}_1^*\mathbf{A}_1\right)_{.i}\left(\widehat{\mathbf{c}}_{.j}^{(1)}\right)\right)_\beta^\beta}{\sum_{\beta\in J_{r_1,n}}\left|\mathbf{A}_1^*\mathbf{A}_1\right|_\beta^\beta} + \frac{\sum_{\beta\in J_{r_3,n}\{i\}}\mathrm{cdet}_i\left(\left(\mathbf{H}^*\mathbf{H}\right)_{.i}\left(\widehat{\mathbf{c}}_{.j}^{(2)}\right)\right)_\beta^\beta}{\sum_{\beta\in J_{r_3,n}}\left|\mathbf{H}^*\mathbf{H}\right|_\beta^\beta}$$
$$-\sum_{l=1}^{n}\frac{\sum_{\beta\in J_{r_3,n}\{i\}}\mathrm{cdet}_i\left(\left(\mathbf{H}^*\mathbf{H}\right)_{.i}\left(\widehat{\mathbf{a}}_{.l}^{(2)}\right)\right)_\beta^\beta}{\sum_{\beta\in J_{r_3,n}}\left|\mathbf{H}^*\mathbf{H}\right|_\beta^\beta} \cdot \frac{\sum_{\beta\in J_{r_1,n}\{l\}}\mathrm{cdet}_l\left(\left(\mathbf{A}_1^*\mathbf{A}_1\right)_{.l}\left(\widehat{\mathbf{c}}_{.j}^{(1)}\right)\right)_\beta^\beta}{\sum_{\beta\in J_{r_1,n}}\left|\mathbf{A}_1^*\mathbf{A}_1\right|_\beta^\beta},\tag{41}$$

where $\widehat{\mathbf{c}}_{.j}^{(1)}$, $\widehat{\mathbf{c}}_{.j}^{(2)}$, and $\widehat{\mathbf{a}}_{.j}^{(2)}$ are the jth columns of the matrices $\widehat{\mathbf{C}}_1$, $\widehat{\mathbf{C}}_2$, and $\widehat{\mathbf{A}}_2$, respectively.

Proof. The proof is similar to the proof of Theorem 3.1.

5. Conclusion

Within the framework of the theory of row-column determinants previously introduced by the author, we get determinantal representations (analogs of Cramer's rule) of partial solutions to the system of two-sided quaternion matrix equations $\mathbf{A}_1\mathbf{X}\mathbf{B}_1=\mathbf{C}_1$, $\mathbf{A}_2\mathbf{X}\mathbf{B}_2=\mathbf{C}_2$, and its special cases with 1 and 2 one-sided matrix equations. We use previously obtained by the author determinantal representations of the Moore-Penrose inverse. Note to give determinantal representations for all above matrix systems over the complex field, it is obviously needed to substitute all row and column determinants by usual determinants.

Conflict of interest

The author declares that there are no conflict interests.

Author details

Ivan I. Kyrchei

Address all correspondence to: kyrchei@online.ua

Pidstrygach Institute for Applied Problems of Mechanics and Mathematics, NAS of Ukraine, Lviv, Ukraine

References

[1] Mitra SK. A pair of simultaneous linear matrix $A_1XB_1=C_1$ and $A_2XB_2=C_2$. Proceedings of the Cambridge Philosophical Society. 1973;**74**:213-216

[2] Mitra SK. A pair of simultaneous linear matrix equations and a matrix programming problem. Linear Algebra and its Applications. 1990;**131**:97-123. DOI: 10.1016/0024-3795(90)90377-O

[3] Shinozaki N, Sibuya M. Consistency of a pair of matrix equations with an application. Keio Engineering Report. 1974;**27**:141-146

[4] Van der Woulde J. Feedback decoupling and stabilization for linear system with multiple exogenous variables [PhD thesis]. Netherlands: Technical University of Eindhoven; 1987

[5] Özgüler AB, Akar N. A common solution to a pair of linear matrix equations over a principle domain. Linear Algebra and its Applications. 1991;**144**:85-99. DOI: 10.1016/0024-3795(91)90063-3

[6] Navarra A, Odell PL, Young DM. A representation of the general common solution to the matrix equations $A_1XB_1=C_1$ and $A_2XB_2=C_2$ with applications. Computers & Mathematics with Applications. 2001;**41**:929-935. DOI: 10.1016/S0898-1221(00)00330-8

[7] Wang QW. The general solution to a system of real quaternion matrix equations. Computers & Mathematics with Applications. 2005;**49**:665-675. DOI: 10.1016/j.camwa.2004.12.00

[8] Kyrchei I. Cramer's rule for generalized inverse solutions. In: Kyrchei I, editor. Advances in Linear Algebra Research. New York: Nova Sci. Publ; 2015. pp. 79-132

[9] Aslaksen H. Quaternionic determinants. Mathematical Intelligence. 1996;**18**(3):57-65. DOI: 10.1007/BF03024312

[10] Cohen N, De Leo S. The quaternionic determinant. Electronic Journal of Linear Algebra. 2000;**7**:100-111. DOI: 10.13001/1081-3810.1050

[11] Dieudonne J. Les determinants sur un corps non-commutatif. Bulletin de la Société Mathématique de France. 1943;**71**:27-45

[12] Study E. Zur theorie der linearen gleichungen. Acta Mathematica. 1920;**42**:1-61

[13] Cayley A. On certain results relating to quaternions. Philosophical Magazine. 1845;**26**:141-145. Reprinted in The collected mathematical papers. Cambridge Univ. Press. 1889;**1**:123-126

[14] Moore EH. On the determinant of an hermitian matrix of quaternionic elements. Bulletin of the American Mathematical Society. 1922;**28**:161-162

[15] Dyson FJ. Quaternion determinants. Helvetica Physica Acta. 1972;**45**:289-302. DOI: 10.5169/seals-114,385

[16] Chen L. Definition of determinant and Cramer solution over the quaternion field. Acta Mathematica Sinica. 1991;**7**:171-180. DOI: 10.1007/BF02633946

[17] Kyrchei I. Cramer's rule for quaternion systems of linear equations. Fundamentalnaya i Prikladnaya Matematika. 2007;**13**(4):67-94

[18] Kyrchei I. The theory of the column and row determinants in a quaternion linear algebra. In: Baswell AR, editor. Advances in Mathematics Research 15. New York: Nova Sci. Publ; 2012. pp. 301-359

[19] Kyrchei I. Determinantal representations of the Moore-Penrose inverse over the quaternion skew field and corresponding Cramer's rules. Linear Multilinear Algebra. 2011;**59**(4):413-431. DOI: 10.1080/03081081003586860

[20] Kyrchei I. Explicit representation formulas for the minimum norm least squares solutions of some quaternion matrix equations. Linear Algebra and its Applications. 2013;**138**(1):136-152. DOI: 10.1016/j.laa.2012.07.049

[21] Kyrchei I. Determinantal representations of the Drazin inverse over the quaternion skew field with applications to some matrix equations. Applied Mathematics and Computation. 2014;**238**:193-207. DOI: 10.1016/j.amc.2014.03.125

[22] Kyrchei I. Determinantal representations of the W-weighted Drazin inverse over the quaternion skew field. Applied Mathematics and Computation. 2015;**264**:453-465. DOI: 10.1016/j.amc.2015.04.125

[23] Kyrchei I. Explicit determinantal representation formulas of W-weighted Drazin inverse solutions of some matrix equations over the quaternion skew field. Mathematical Problems in Engineering. 2016. 13 p. DOI: 10.1155/2016/8673809; ID 8673809

[24] Kyrchei I. Determinantal representations of the Drazin and W-weighted Drazin inverses over the quaternion skew field with applications. In: Griffin S, editor. Quaternions: Theory and Applications. New York: Nova Sci. Publ; 2017. pp. 201-275

[25] Kyrchei I. Weighted singular value decomposition and determinantal representations of the quaternion weighted Moore-Penrose inverse. Applied Mathematics and Computation. 2017;**309**:1-16. DOI: 10.1016/j.amc.2017.03.048

[26] Song GJ, Wang QW, Chang HX. Cramer rule for the unique solution of restricted matrix equations over the quaternion skew field. Computers & Mathematcs with Applications. 2011;**61**:1576-1589. DOI: 10.1016/j.camwa.2011.01.026

[27] Song GJ, Dong CZ. New results on condensed Cramer's rule for the general solution to some restricted quaternion matrix equations. Journal of Applied Mathematics and Computing. 2017;**53**:321-341. DOI: 10.1007/s12190-015-0970-y

[28] Song GJ, Wang QW. Condensed Cramer rule for some restricted quaternion linear equations. Applied Mathematics and Computation. 2011;**218**:3110-3121. DOI: 10.1016/j.amc.2011.08.038

[29] Song G. Characterization of the W-weighted Drazin inverse over the quaternion skew field with applications. Electronic Journal of Linear Algebra. 2013;**26**:1-14. DOI: 10.13001/1081-3810.1635

[30] Wang QW. A system of matrix equations and a linear matrix equation over arbitrary regular rings with identity. Linear Algebra and its Applications. 2004;**384**:43-54. DOI: 10.1016/j.laa.2003.12.039

[31] Wang QW, Wu ZC, Lin CY. Extremal ranks of a quaternion matrix expression subject to consistent systems of quaternion matrix equations with applications. Applied Mathematics and Computation. 2006;**182**(2):1755-1764. DOI: 10.1016/j.amc.2006.06.012

[32] Maciejewski AA, Klein CA. Obstacle avoidance for kinematically redundant manipulators in dynamically varying environments. The International Journal of Robotics Research. 1985; **4**(3):109-117. DOI: 10.1177/027836498500400308

Chapter 2

Matrices Which are Discrete Versions of Linear Operations

Armando Martínez Pérez and Gabino Torres Vega

Additional information is available at the end of the chapter

http://dx.doi.org/10.5772/intechopen.74356

Abstract

We introduce and study a matrix which has the exponential function as one of its eigenvectors. We realize that this matrix represents a set of finite differences derivation of vectors on a partition. This matrix leads to new expressions for finite differences derivatives which are exact for the exponential function. We find some properties of this matrix, the induced derivatives and of its inverse. We provide an expression for the derivative of a product, of a ratio, of the inverse of vectors, and we also find the equivalent of the summation by parts theorem of continuous functions. This matrix could be of interest to discrete quantum mechanics theory.

Keywords: exact finite differences derivative, exact derivatives on partitions, exponential function on a partition, discrete quantum mechanics

1. Introduction

We are interested on matrices which are a local, as well as a global, exact discrete representation of operations on functions of continuous variable, so that there is congruency between the continuous and the discrete operations and properties of functions. Usual finite difference methods [1–4] become exact only in the limit of zero separation between the points of the mesh. Here, we are interested in having exact representations of operations and functions for *finite* separation between mesh points.

The difference between our method and the usual finite differences method is the quantity that appears in the denominator of the definition of derivative. The appropriate choice of that denominator makes possible that the finite differences expressions for the derivative gives the exact results for the exponential function. We concentrate on the derivative operation, and we define a matrix which represents the exact finite difference derivation on a local and a global

© 2018 The Author(s). Licensee IntechOpen. This chapter is distributed under the terms of the Creative Commons Attribution License (http://creativecommons.org/licenses/by/3.0), which permits unrestricted use, distribution, and reproduction in any medium, provided the original work is properly cited. (cc) BY

scale. The inverse of this matrix is just the integration operation. These are interesting subjects by itself, but they are also of interest in the quantum physics realm [5–7].

In this chapter, we will consider only the case of the derivative and the integration of the exponential function.

2. A matrix with the exponential function as an eigenvector

Here, we consider the $N \times N$ antisymmetric, tridiagonal matrix

$$
\mathbf{D}_N := \begin{pmatrix}
\frac{-e^{-v\Delta}}{2\chi(v,\Delta)} & \frac{1}{2\chi(v,\Delta)} & 0 & \cdots & 0 & 0 & 0 \\
\frac{-1}{2\chi(v,\Delta)} & 0 & \frac{1}{2\chi(v,\Delta)} & \cdots & 0 & 0 & 0 \\
0 & \frac{-1}{2\chi(v,\Delta)} & 0 & \cdots & 0 & 0 & 0 \\
\vdots & & & & & & \\
0 & 0 & 0 & \cdots & 0 & \frac{1}{2\chi(v,\Delta)} & 0 \\
0 & 0 & 0 & \cdots & \frac{-1}{2\chi(v,\Delta)} & 0 & \frac{1}{2\chi(v,\Delta)} \\
0 & 0 & 0 & \cdots & 0 & \frac{-1}{2\chi(v,\Delta)} & \frac{e^{v\Delta}}{2\chi(v,\Delta)}
\end{pmatrix}, \tag{1}
$$

where $v \in \mathbb{C}$—it can be pure real or pure imaginary—, $\Delta \in \mathbb{R}^+$, and $\chi(v,\Delta) := \sinh(v\Delta)/v \approx \Delta + v^2\Delta^3/6 + O(\Delta^5)$. This function $\chi(v,\Delta)$ is well defined for $v = 0$, with value $\chi(0,\Delta) = \Delta$. This matrix is interesting because, as we will see below, it represents a derivation on a partition. A rescaled matrix $\overline{\mathbf{D}}_N$ is defined as

$$
\overline{\mathbf{D}}_N := \begin{pmatrix}
-1/z & 1 & 0 & \cdots & 0 & 0 & 0 \\
-1 & 0 & 1 & \cdots & 0 & 0 & 0 \\
0 & -1 & 0 & \cdots & 0 & 0 & 0 \\
\vdots & & & & & & \\
0 & 0 & 0 & \cdots & 0 & 1 & 0 \\
0 & 0 & 0 & \cdots & -1 & 0 & 1 \\
0 & 0 & 0 & \cdots & 0 & -1 & z
\end{pmatrix}, \tag{2}
$$

where $z = e^{v\Delta}$, and

$$
\mathbf{D}_N := \frac{\overline{\mathbf{D}}_N}{2\chi(v,\Delta)}. \tag{3}
$$

We are mainly interested in finding the eigenvalues and the corresponding eigenvectors of these matrices.

We start our study with a result about the determinant of $\overline{\mathbf{D}}_N - \lambda \mathbf{I}_N$,

$$|\overline{\mathbf{D}}_N - \lambda \mathbf{I}_N| = |\overline{\mathbf{D}}_N + \alpha \mathbf{I}_N|$$

$$= \begin{vmatrix} \alpha - 1/z & 1 & 0 & 0 & \ldots & 0 & 0 & 0 \\ -1 & \alpha & 1 & 0 & \ldots & 0 & 0 & 0 \\ 0 & -1 & \alpha & 1 & \ldots & 0 & 0 & 0 \\ 0 & 0 & -1 & \alpha & \ldots & 0 & 0 & 0 \\ & \vdots & & & & & & \\ 0 & 0 & 0 & \ldots & \alpha & 1 & 0 & 0 \\ 0 & 0 & 0 & \ldots & -1 & \alpha & 1 & 0 \\ 0 & 0 & 0 & \ldots & 0 & -1 & \alpha & 1 \\ 0 & 0 & 0 & \ldots & 0 & 0 & -1 & \alpha + z \end{vmatrix} \tag{4}$$

$$= \left(\alpha - \frac{1}{z} \right) A_{N-1}(\alpha) + A_{N-2}(\alpha),$$

where $\lambda = -\alpha$,

$$A_j(\alpha) := \begin{vmatrix} \alpha & 1 & 0 & \ldots & 0 & 0 & 0 \\ -1 & \alpha & 1 & \ldots & 0 & 0 & 0 \\ 0 & -1 & \alpha & \ldots & 0 & 0 & 0 \\ & \vdots & & & & & \\ 0 & 0 & \ldots & \alpha & 1 & 0 & 0 \\ 0 & 0 & \ldots & -1 & \alpha & 1 & 0 \\ 0 & 0 & \ldots & 0 & -1 & \alpha & 1 \\ 0 & 0 & \ldots & 0 & 0 & -1 & \alpha + z \end{vmatrix} \tag{5}$$

$$= (\alpha + z) B_{j-1}(\alpha) + B_{j-2}(\alpha),$$

and

$$B_j(\alpha) = \begin{vmatrix} \alpha & 1 & 0 & \ldots & 0 & 0 \\ -1 & \alpha & 1 & \ldots & 0 & 0 \\ 0 & -1 & \alpha & \ldots & 0 & 0 \\ & \vdots & & & & \\ 0 & 0 & \ldots & \alpha & 1 & 0 \\ 0 & 0 & \ldots & -1 & \alpha & 1 \\ 0 & 0 & \ldots & 0 & -1 & \alpha \end{vmatrix}. \tag{6}$$

Strikingly, we recognize the determinant $B_j(\alpha)$ as the Fibonacci polynomial of index $j + 1$ [10, 11], i.e., $B_j(\alpha) = F_{j+1}(\alpha)$. Fibonacci polynomials are defined as

$$F_0(x) = 0, \quad F_1(x) = 1, \quad F_j(x) = xF_{j-1}(x) + F_{j-2}(x), \quad j \geq 2. \tag{7}$$

Since we have that $B_j(\alpha) = F_{j+1}(\alpha)$, and the recursion relationship for Fibonacci polynomials, we also have that

$$A_j(\alpha) = (\alpha + z)F_j(\alpha) + F_{j-1}(\alpha) = zF_j(\alpha) + F_{j+1}(\alpha), \tag{8}$$

and then

$$
\begin{aligned}
&|\mathbf{D}_N + \alpha \mathbf{I}_N| \\
&= \left(\alpha - \frac{1}{z}\right)[zF_{N-1}(\alpha) + F_N(\alpha)] + zF_{N-2}(\alpha) + F_{N-1}(\alpha) \\
&= z[\alpha F_{N-1}(\alpha) + F_{N-2}(\alpha)] + \left(\alpha - \frac{1}{z}\right)F_N(\alpha) \\
&= \left(\alpha + z - \frac{1}{z}\right)F_N(\alpha).
\end{aligned}
\tag{9}
$$

Then, the eigenvalues of the derivative matrix $\overline{\mathbf{D}}_N$ are $\lambda_1 = z - 1/z = e^{v\Delta} - e^{-v\Delta} = 2\sinh(v\Delta)$ and $\lambda_m = -\alpha_m$, where α_m is the m-th root of the N-th Fibonacci polynomial, which is a polynomial of degree $N - 1$ [10, 11].

The system of simultaneous equations for the eigenvector $e_m^T = (e_{m,1} e_{m,2}, ..., e_N)$ corresponding to λ_m, can be put in a form similar to the recursion relationship for the Fibonacci polynomials, i.e.,

$$e_{m,2} = \lambda_m e_{m,1} + \frac{e_{m,1}}{z}, \tag{10}$$

$$e_{m,j+1} = \lambda_m e_{m,j} + e_{m,j-1}, \quad 1 < j < N, \tag{11}$$

$$z e_{m,N} = \lambda_m e_{m,N} + e_{m,N-1}. \tag{12}$$

This set of recursion relationships can be written as the matrix equation

$$\begin{pmatrix} e_{m,j} \\ e_{m,j+1} \end{pmatrix} = \begin{pmatrix} 0 & 1 \\ 1 & \lambda_m \end{pmatrix} \begin{pmatrix} e_{m,j-1} \\ e_{m,j} \end{pmatrix}, \quad j = 1, ..., N, \tag{13}$$

where $e_{m,0} = e_{m,1}/z$ and $e_{m,N+1} = z e_{m,N}$. Thus

$$\begin{pmatrix} e_{m,j} \\ e_{m,j+1} \end{pmatrix} = \begin{pmatrix} 0 & 1 \\ 1 & \lambda_m \end{pmatrix}^j \begin{pmatrix} e_{m,0} \\ e_{m,1} \end{pmatrix}, \quad j = 1, ..., N, \tag{14}$$

but

$$\begin{pmatrix} 0 & 1 \\ 1 & \lambda_m \end{pmatrix}^j = \begin{pmatrix} F_{j-1}(\lambda_m) & F_j(\lambda_m) \\ F_j(\lambda_m) & F_{j+1}(\lambda_m) \end{pmatrix}, \tag{15}$$

and then,

$$\begin{pmatrix} e_{m,j} \\ e_{m,j+1} \end{pmatrix} = \begin{pmatrix} F_{j-1}(\lambda_m) & F_j(\lambda_m) \\ F_j(\lambda_m) & F_{j+1}(\lambda_m) \end{pmatrix} \begin{pmatrix} e_{m,0} \\ e_{m,1} \end{pmatrix}, \quad j = 1, \dots, N. \tag{16}$$

i.e., the j-th component of the m-th eigenvector is

$$e_{m,j} = \left[F_j(\lambda_m) + \frac{F_{j-1}(\lambda_m)}{z} \right] e_{m,1} \text{ for } j = 1, 2, \dots, N. \tag{17}$$

For the case of the eigenvalue $\lambda_1 = z - 1/z$, we can rewrite Eq. (17) by noticing that if we let $x = w - w^{-1}$ ($w \in \mathbb{C}$), then $F_n(x) + F_{n-1}(x)/w = w^{n-1}$ for $n = 1, 2, \dots$. This can be proved by induction method as follows. For $n = 1$, it is immediately verified. First, suppose that the equality holds for $n \le k$. Next, we compute the right-hand side of the equality for $k + 1$. Substituting $F_{k-1} = w(w^{k-1} - F_k)$ in the expression for $k + 1$, and using the properties of the Fibonacci polynomials, we obtain

$$F_{k+1}(x) + \frac{F_k(x)}{w} = x F_k(x) + F_{k-1}(x) + \frac{F_k(x)}{w}$$
$$= x F_k(x) + w^k - w F_k(x) + \frac{F_k(x)}{w} \tag{18}$$
$$= w^k.$$

Therefore, according to Eqs. (17) and (18), the eigenvector for the eigenvalue $\lambda_1 = 2 \sinh(v\Delta)$ takes the form $\mathbf{e}_1 = c(1, z, \dots, z^{N-1})^T$, where c is a normalization constant. We can take advantage of the normalization constant and write

$$\mathbf{e}_1 = c(e^{vq_1}, , e^{vq_2}, \dots, , e^{vq_N})^T, \tag{19}$$

with eigenvalue $\lambda_1 = v$ (in original scaling, i.e., the eigenvalue of the matrix $\mathbf{D}_N$), q_1 is an arbitrary constant, and $q_j = q_1 + (j-1)\Delta$. This means that the exponential function is an eigenvector of the derivative matrix which is a global representation of the derivative on the partition $\{q_1, q_2, \dots, q_N\}$. Recall that the exponential function is an eigenfunction of the derivative of functions of continuous variable.

The remain of the eigenvectors have eigenvalues equal to the negative of the roots of the N-th Fibonacci polynomial $\lambda_m = -x_m$, $m = 1, 2, \dots, N - 1$, and have the form

$$\mathbf{e}_m = c \begin{pmatrix} 1 \\ F_2(\lambda_m) + e^{-v\Delta} \\ F_3(\lambda_m) + e^{-v\Delta} F_2(\lambda_m) \\ \vdots \\ F_{N-1}(\lambda_m) + e^{-v\Delta} F_{N-2}(\lambda_m) \\ e^{-v\Delta} F_{N-1}(\lambda_m) \end{pmatrix} \tag{20}$$

The vector that we will be interested on is the one which is the exponential function (19) with eigenvalue v.

3. The matrix $\mathbf{D}_N$ represents a derivation

Let us consider a partition, $P(N):=\{q_i\}_1^N$, $q_i \in \mathbb{R}$, of N equally spaced points q_i of the interval $[a, b] \in \mathbb{R}$, $a < b$, with the same separation $\Delta = (b - a)/(N - 1)$ between them.

The rows of the result of the multiplication of the derivative matrix $\mathbf{D}_N$ and a vector $\mathbf{g}:=(g_1, g_2, \ldots, g_n)^T$ are

$$(\mathbf{D}_N \mathbf{g})_j = \frac{g_{j+1} - g_{j-1}}{2\chi(v, \Delta)}, \quad j = 1, 2, \ldots, N, \tag{21}$$

where $g_0:=e^{-v\Delta} g_1$ and $g_{N+1}:=e^{v\Delta} g_N$. We recognize these expressions as the second order derivatives of the function $g(x)$ at the mesh points, but instead of dividing by twice the separation Δ between the mesh points, there is the function $\chi(v, \Delta)$ in the denominator. This function makes it possible that the exponential function be an eigenvector of the matrix $\mathbf{D}_N$.

The values $g_0 = e^{-v\Delta} g_1$ and $g_{N+1} = e^{v\Delta} g_N$ extend the original interval $[a, b]$ to $[a - \Delta, b + \Delta]$ so that we have well defined the second order derivatives at all the points of the initial partition, including the edges of the interval. When $g(x)$ is the exponential function, we have $g_0 = e^{v(x_1 - \Delta)}$ and $g_{N+1} = e^{v(x_N + \Delta)}$, i.e., they are the values of the exponential function evaluated at the points of the extension.

Thus, we define finite differences derivatives for any function $g(x)$ defined on the partition as

$$(Dg)_1 = \frac{g_2 - e^{-v\Delta} g_1}{2\chi(v, \Delta)}, \tag{22}$$

$$(Dg)_j = \frac{g_{j+1} - g_{j-1}}{2\chi(v, \Delta)}, \tag{23}$$

$$(Dg)_N = \frac{e^{v\Delta} g_N - g_{N-1}}{2\chi(v, \Delta)}, \tag{24}$$

to be used on the first, central, and last points of the partition.

The determinant of the derivative matrix is not always zero, and in fact, it is [see Eqs. (4) and (9)]

$$|\overline{\mathbf{D}}_N| = 2\sinh(v\Delta) F_N(0). \tag{25}$$

But, since $F_{2j+1} = 1$, and $F_{2j} = 0$, then

$$|\overline{\mathbf{D}}_{2j}| = 0, \quad |\overline{\mathbf{D}}_{2j+1}| = 2\sinh(v\Delta). \tag{26}$$

Hence, only the matrices with an odd dimension have an inverse.

Next, we will derive some properties of these finite differences derivatives.

3.1. The derivative of a product of vectors

There are two equivalent expressions for the finite differences derivative of a product of vectors defined on the partition. A set of such expressions is

$$(Dgh)_1 = \frac{g_2 h_2 - e^{-v\Delta} g_1 h_1}{2\chi(v, \Delta)}$$

$$= \frac{g_2 h_2 - e^{-v\Delta} g_1 h_2}{2\chi(v, \Delta)} + g_1 \frac{e^{-v\Delta} h_2 - h_2 + h_2 - e^{-v\Delta} h_1}{2\chi(v, \Delta)}$$

$$\tag{27}$$

$$= h_2 (Dg)_1 + g_1 (Dh)_1 + g_1 h_2 \frac{e^{-v\Delta} - 1}{2\chi(v, \Delta)}$$

$$= h_2 (Dg)_1 + g_1 (Dh)_1 + g_1 h_2 \left[-\frac{v}{2} + \frac{v^2}{4}\Delta + O(\Delta^3) \right],$$

$$(Dgh)_j = h_{j+1} (Dg)_j + g_{j-1} (Dh)_j, \tag{28}$$

$$(Dgh)_N = h_N (Dg)_N + g_{N-1} (Dh)_N + \frac{1 - e^{v\Delta}}{2\chi(v, \Delta)} g_{N-1} h_N$$

$$\tag{29}$$

$$\approx h_N (Dg)_N + g_{N-1} (Dh)_N + g_{N-1} h_N \left[-\frac{v}{2} - \frac{v^2}{4}\Delta + O(\Delta^3) \right].$$

A second set of equalities is

$$(Dgh)_1 = g_2 (Dh)_1 + h_1 (Dg)_1 + g_2 h_1 \frac{e^{-v\Delta} - 1}{2\chi(v, \Delta)}$$

$$\tag{30}$$

$$= g_2 (Dh)_1 + h_1 (Dg)_1 + g_2 h_1 \left[-\frac{v}{2} + \frac{v^2}{4}\Delta + O(\Delta^3) \right],$$

$$(Dgh)_j = g_{j+1} (Dh)_j + h_{j-1} (Dg)_j, \tag{31}$$

$$(Dgh)_N = g_N (Dh)_N + h_{N-1} (Dg)_N + g_N h_{N-1} \frac{1 - e^{v\Delta}}{2\chi(v, \Delta)}$$

$$\tag{32}$$

$$\approx g_N (Dh)_N + h_{N-1} (Dg)_N + g_N h_{N-1} \left[-\frac{v}{2} - \frac{v^2}{4}\Delta + O(\Delta^3) \right],$$

3.2. Summation by parts

The sum of Eqs. (28) or (31), with weights $2\chi(v, \Delta)$, results in

$$
\sum_{j=n}^{m} 2\chi(v, \Delta)h_{j+1}(Dg)_j + \sum_{j=n}^{m} 2\chi(v, \Delta)g_{j-1}(Dh)_j
$$

$$
= \sum_{j=n}^{m} 2\chi(v, \Delta)(Dgh)_j \tag{33}
$$

$$
= g_{m+1}h_{m+1} + g_m h_m - g_n h_n - g_{n-1}h_{n-1},
$$

or

$$
\sum_{j=n}^{m} 2\chi(v, \Delta)g_{j+1}(Dh)_j + \sum_{j=n}^{m} 2\chi(v, \Delta)h_{j-1}(Dg)_j \tag{34}
$$

$$
= g_{m+1}h_{m+1} + g_m h_m - g_n h_n - g_{n-1}h_{n-1}.
$$

This is the discrete version of the integration by parts theorem for continuous variable functions, a very useful result.

3.3. Second derivatives

Expressions for higher order derivatives are obtained through the powers of $\mathbf{D}_N$. For instance, for the first two points, the second derivative is

$$
(D^2 g)_1 = \frac{(e^{-2v\Delta} - 1)g_1 - e^{-v\Delta}g_2 + g_3}{4\chi^2(v, \Delta)} = \frac{(Dg)_2 - e^{-v\Delta}(Dg)_1}{2\chi(v, \Delta)}, \tag{35}
$$

$$
(D^2 g)_2 = \frac{e^{-v\Delta}g_1 - 2g_2 + g_4}{4\chi^2(v, \Delta)} = \frac{(Dg)_3 - (Dg)_1}{2\chi(v, \Delta)}, \tag{36}
$$

For inner points we get

$$
(D^2 g)_j = \frac{g_{j-2} - 2g_j + g_{j+2}}{4\chi^2(v, \Delta)} = \frac{(Dg)_{j+1} - (Dg)_{j-1}}{2\chi(v, \Delta)}, \quad 3 \leq j \leq N - 3, \tag{37}
$$

and for the last two points of the mesh, we find

$$
(D^2 g)_{N-1} = \frac{g_{N-3} - 2g_{N-1} + e^{v\Delta}g_N}{4\chi^2(v, \Delta)} = \frac{(Dg)_N - (Dg)_{N-2}}{2\chi(v, \Delta)}, \tag{38}
$$

$$
(D^2 g)_N = \frac{g_{N-2} - e^{v\Delta}g_{N-1} + (e^{2v\Delta} - 1)g_N}{4\chi^2(v, \Delta)}
$$

$$
= \frac{e^{v\Delta}(Dg)_N - (Dg)_{N-1}}{\chi_2(v, 2\Delta)}. \tag{39}
$$

These derivatives also have the exponential function as one of their eigenvectors, and we can generate expressions for higher derivatives with higher powers of the derivative matrix.

3.4. The derivative of the inverse of functions

It is possible to give an expression for the derivative of $h^{-1}(q)$, including the edge points. For the first point, we have

$$
\begin{aligned}
\left(D\frac{1}{h} \right)_1 &= \frac{1}{2\chi(v,\Delta)} \left(\frac{1}{h_2} - \frac{e^{-v\Delta}}{h_1} \right) \\
&= \frac{1}{2\chi(v,\Delta)} \left(-\frac{h_2 - h_1}{h_1 h_2} + \frac{1 - e^{-v\Delta}}{h_1} \right) \\
&= -\frac{(Dh)_1}{h_1 h_2} + \frac{1 - e^{-v\Delta}}{2\chi(v,\Delta)} \left(\frac{1}{h_1} + \frac{1}{h_2} \right).
\end{aligned}
\tag{40}
$$

For central and last points, we find that

$$
\left(D\frac{1}{h} \right)_j = -\frac{(Dh)_j}{h_{j-1} h_{j+1}},
\tag{41}
$$

$$
\left(D\frac{1}{h} \right)_N = -\frac{(Dh)_N}{h_{N-1} h_N} + \frac{e^{v\Delta} - 1}{2\chi(v,\Delta)} \left(\frac{1}{h_{N-1}} + \frac{1}{h_N} \right).
\tag{42}
$$

The derivatives for the first and last points coincide with the derivative for central points when $\Delta = 0$.

3.5. The derivative of the ratio of functions

Now, we take advantage of the derivative for the inverse of a function and the derivative of a product of functions and obtain what the derivative of a ratio of functions is

$$
\begin{aligned}
\left(D\frac{g}{h} \right)_1 &= \frac{1}{h_2} (Dg)_1 + g_1 \left(D\frac{1}{h} \right)_1 + \frac{g_1}{h_2} \frac{e^{-v\Delta} - 1}{2\chi(v,\Delta)} \\
&= \frac{1}{h_2} (Dg)_1 + g_1 \left[-\frac{(Dh)_1}{h_1 h_2} + \frac{1}{2\chi(v,\Delta)} \left(\frac{1}{h_1} + \frac{1 - e^{-v\Delta}}{h_2} \right) \right] + \frac{g_1}{h_2} \frac{e^{-v\Delta} - 1}{2\chi(v,\Delta)} \\
&= \frac{1}{h_2} (Dg)_1 - \frac{g_1}{h_1 h_2} (Dh)_1 + \frac{g_1}{h_1} \frac{1 - e^{-v\Delta}}{2\chi(v,\Delta)},
\end{aligned}
\tag{43}
$$

$$
\left(D\frac{g}{h} \right)_j = \frac{(Dg)_j}{h_{j-1}} - g_{j+1} \frac{(Dh)_j}{h_{j+1} h_{j-1}},
\tag{44}
$$

$$
\left(D\frac{g}{h} \right)_N = \frac{1}{h_N} (Dg)_N - \frac{g_{N-1}}{h_{N-1} h_N} (Dh)_N + \frac{g_{N-1}}{h_{N-1}} \frac{e^{v\Delta} - 1}{2\chi(v,\Delta)}.
\tag{45}
$$

expressions which are very similar to the continuous variable results. Again, these expressions coincide in the limit $\Delta \to 0$, and they reduce to the corresponding expressions for continuous variables.

3.6. The local inverse operation of the derivative

The inverse operation to the finite differences derivative, at a given point, is the summation with weights $2\chi(v, \Delta)$

$$\sum_{j=n}^{m} 2\chi(v, \Delta)(Dg)_j = \sum_{j=n}^{m} \left(g_{j+1} - g_{j-1} \right) = g_{m+1} + g_m - g_n - g_{n-1}. \tag{46}$$

This equality is the equivalent to the usual result for continuous functions, $\int_a^x dy(dg(y)/dy) = g(x) - g(a)$. Note that the inverse at the local level is a bit different from the expressions obtained by means of the inverse matrix $\mathbf{S}$ (see below) of the derivative matrix $\mathbf{D}$. When dealing with matrices there are no boundary terms to worry about.

3.7. An eigenfunction of the summation operation

Because the exponential function is an eigenfunction of the finite differences derivative and according to Eq. (46), we can say that

$$\sum_{j=n}^{m} 2\chi(v, \Delta)ve^{vq_j} = \sum_{j=n}^{m} 2\chi(v, \Delta)(De^{vq})_j = \sum_{j=n}^{m}\left(e^{vq_{j+1}} - e^{vq_{j-1}}\right)$$

$$= e^{vq_{m+1}} + e^{vq_m} - e^{vq_n} - e^{vq_{n-1}}, \tag{47}$$

in agreement with the corresponding continuous variable equality $\int_a^x dx v e^{vx} = e^{vx} - e^{va}$. However, here, we have to deal with two values at each boundary.

3.8. The chain rule

The chain rule also has a finite differences version. That version is

$$(Dg(h(q)))_j = \frac{g\left(h\left(q_{j+1}\right)\right) - g\left(h\left(q_{j-1}\right)\right)}{2\chi(v, \Delta)}$$

$$= \frac{g\left(h\left(q_{j+1}\right)\right) - g\left(h\left(q_{j-1}\right)\right)}{2\chi\left(v, h\left(q_{j+1}\right) - h\left(q_j\right)\right)} \frac{2\chi\left(v, h\left(q_{j+1}\right) - h\left(q_j\right)\right)}{2\chi(v, \Delta)} \tag{48}$$

$$= (Dg(h))_j \frac{\chi\left(v, h\left(q_{j+1}\right) - h\left(q_j\right)\right)}{\chi(v, \Delta)}$$

where

$$(Dg(h))_j := \frac{g\left(h\left(q_{j+1}\right)\right) - g\left(h\left(q_{j-1}\right)\right)}{2\chi\left(v, h\left(q_{j+1}\right) - h\left(q_j\right)\right)} \tag{49}$$

is a finite differences derivative of $g(h)$ with respect to h, and the second factor approaches the derivative of $h(q)$ with respect to q

$$\frac{\chi\left(v, h\left(q_{j+1}\right) - h\left(q_j\right)\right)}{\chi(v, \Delta)} \approx \frac{h\left(q_{j+1}\right) - h\left(q_j\right) + O(\Delta h^2)}{\Delta + O(\Delta^2)}. \tag{50}$$

Thus, we will recover the usual chain rule for continuous variable functions in the limit $\Delta \to 0$.

4. The commutator between coordinate and derivative

Let us determine the commutator, from a local point of view first, between the coordinate—the points of the partition $P(N)$—and the finite differences derivative. We begin with the derivative of q,

$$(Dq)_j = \frac{q_{j+1} - q_{j-1}}{2\chi(v, \Delta)} = \frac{\Delta}{\chi(v, \Delta)} \approx 1 - \frac{v^2}{6}\Delta^2. \tag{51}$$

Hence, the finite differences derivative of the product $qg(q)$ is

$$(Dqg)_j = q_{j+1}(Dg)_j + g_{j-1}(Dq)_j = q_{j+1}(Dg)_j + g_{j-1}\frac{\Delta}{\chi(v, \Delta)}, \tag{52}$$

i.e.,

$$(D_c qg)_j - q_{j+1}(D_c g)_j = g_{j-1}\frac{\Delta}{\chi(v, \Delta)}. \tag{53}$$

This is the finite differences version of the commutator between the coordinate q and the finite differences derivative D. This equality will become the identity operator in the small Δ limit, as expected. An equivalent expression is

$$(Dqg)_j - q_{j-1}(Dg)_j = g_{j+1}\frac{\Delta}{\chi(v, \Delta)}. \tag{54}$$

This is the finite differences version of the commutator between coordinate and derivative; the right hand side of this equality becomes g_j in the small Δ limit, i.e., it becomes the identity operator.

4.1. The commutator between the derivative and coordinate matrices

The commutator between the partition and the finite differences derivative can also be calculated from a global point of view using the corresponding matrices. Let the diagonal matrix $[\mathbf{Q}_N]$ which will represent the coordinate partition

$$\mathbf{Q}_N := \text{diag}(q_1, q_2, \ldots, q_N). \tag{55}$$

Then, the commutator between the derivative matrix and the coordinate matrix is

$$[\mathbf{D}_N, \mathbf{Q}_N] = \frac{\Delta}{2\chi(v, \Delta)} \begin{pmatrix} 0 & 1 & 0 & 0 & \ldots & 0 & 0 & 0 \\ 1 & 0 & 1 & 0 & \ldots & 0 & 0 & 0 \\ 0 & 1 & 0 & 1 & \ldots & 0 & 0 & 0 \\ & \vdots & & & & & & \\ 0 & 0 & 0 & 0 & \ldots & 0 & 1 & 0 \\ 0 & 0 & 0 & 0 & \ldots & 1 & 0 & 1 \\ 0 & 0 & 0 & 0 & \ldots & 0 & 1 & 0 \end{pmatrix}. \tag{56}$$

This is a kind of nearest neighbors' average operator, inside the interval. The small Δ limit is just

$$[\mathbf{D}_N, \mathbf{Q}_N] \approx I, \tag{57}$$

where $\mathbf{I}$ is the identity matrix, with the first and last elements replace with 1/2. Thus, coordinate and derivative matrices are finite differences conjugate of each other.

5. An integration matrix

Since the determinant of the derivative matrix $\mathbf{D}_N$ is not always zero, we expect that there exist an inverse of it. At a local level, the inverse of the finite differences derivation is the summation as was found in Eq. (46). In this section, we determine the inverse of the derivative matrix, and we find that it is a global finite difference integration operation.

Once we know the eigenvalues and eigenvectors of the derivative matrix $\mathbf{D}_N$, it turns out that we also know the eigenvectors and eigenvalues of the inverse matrix, when it exists. In fact, the equality $\mathbf{D}_N \mathbf{e}_m = \lambda_m \mathbf{e}_m$ with $\lambda_m \neq 0$, imply that

$$\mathbf{D}_N^{-1} \mathbf{e}_m = \lambda_m^{-1} \mathbf{e}_m. \tag{58}$$

The inverse matrix $\mathbf{S}_N = \mathbf{D}_N^{-1}$ is

$$\mathbf{S}_N = \frac{1}{z - \frac{1}{z}} \begin{pmatrix} 1 & -z & 1 & -z & 1 & \cdots & -z & 1 \\ z & -1 & 1/z & -1 & 1/z & \cdots & -1 & 1/z \\ 1 & -1/z & 1 & -z & 1 & \cdots & -z & 1 \\ z & -1 & z & -1 & 1/z & \cdots & -1 & 1/z \\ & \vdots & & & & & & \\ 1 & -1/z & 1 & -1/z & 1 & \cdots & -z & 1 \\ z & -1 & z & -1 & z & \cdots & -1 & 1/z \\ 1 & -1/z & 1 & -1/z & 1 & \cdots & -1/z & 1 \\ z & -1 & z & -1 & z & \cdots & -1 & 1/z \\ 1 & -1/z & 1 & -1/z & 1 & \cdots & -1/z & 1 \end{pmatrix}, \tag{59}$$

Its determinant is

$$|\mathbf{S}_N| = \sinh^{N-1}(v\Delta). \tag{60}$$

This matrix represents an integration on the partition, with an exact value when it is applied to the exponential function e^{vq} on the partition. When applied to an arbitrary vector $\mathbf{g} = (g_1, g_2, \ldots, g_N)^T$, we obtain formulas for the finite differences integration, including the edge points

$$(\mathbf{S}_N\mathbf{g})_1 = \frac{1}{z - 1/z}\left[g_1 + \sum_{i=1}^{M}(g_{2i+1} - zg_{2i})\right], \tag{61}$$

$$(\mathbf{S}_N\mathbf{g})_{2j} = \frac{1}{z - 1/z}\left[zg_1 + \sum_{k=1}^{j-1}(zg_{2k+1} - g_{2k}) + \sum_{k=j}^{M}\left(\frac{g_{2k+1}}{z} - g_{2k}\right)\right], \tag{62}$$

$$(\mathbf{S}_N\mathbf{g})_{2j+1} = \frac{1}{z - 1/z}\left[g_1 + \sum_{k=1}^{j}\left(g_{2k+1} - \frac{g_{2k}}{z}\right) + \sum_{k=j+1}^{M}(g_{2k+1} - zg_{2k})\right], \tag{63}$$

$$(\mathbf{S}_N\mathbf{g})_N = \frac{1}{z - 1/z}\left[g_1 + \sum_{i=1}^{M}\left(g_{2i+1} - \frac{g_{2i}}{z}\right)\right], \tag{64}$$

where $N = 2M + 1$. These are new formulas for discrete integration for the exponential function on a partition of equally separated points with the characteristic that it is exact for the exponential function e^{vq}.

6. Transformation between coordinate and derivative representations

Since one of the eigenvalues of the derivative matrix is a continuous variable, we can talk of conjugate functions with a continuous argument v. The relationship between discrete vectors on a partition $\{q_i\}$ and functions with a continuous argument v makes use of continuous and discrete Fourier type of transformations, a wavelet [12]. If we have a function h of continuous argument v, a conjugate vector on the partition $\{q_i\}$ is defined through the type of continuous Fourier transform F as

$$(Fh)\left(q_j\right) := \frac{1}{L\sqrt{2\Delta}} \int_{-L/2}^{L/2} e^{-iq_j v} h(v) dv, \tag{65}$$

and vice-versa, a continuous variable function is defined with the help of a discrete type of Fourier transform F as

$$(Fg)(v) := \frac{L}{\sqrt{2\Delta}} \sum_{j=-N+1}^{N-1} 2\chi(v,\Delta) e^{iq_j v} g_j. \tag{66}$$

Assuming that the involved integrals converge absolutely, we can say that

$$F(Fg)\left(q_j\right) := \frac{1}{L\sqrt{2\Delta}} \int_{-L/2}^{L/2} e^{-iq_j v} \frac{L}{\sqrt{2\Delta}} \sum_{k=-N+1}^{N-1} 2\chi(v,\Delta) e^{iq_k v} g_k dv$$

$$= \frac{1}{\Delta} \sum_{k=-N+1}^{N-1} g_k \int_{-L/2}^{L/2} e^{i(q_k-q_j)v} \sinh(v\Delta) \frac{dv}{v} \tag{67}$$

$$= \sum_{k=-N+1}^{N-1} g_k K\left(q_k - q_j, L, \Delta\right).$$

where

$$K\left(q_k - q_j, L, \Delta\right) := \frac{1}{\Delta} \int_{-L/2}^{L/2} e^{i(q_k-q_j)v} \sinh(v\Delta) \frac{dv}{v}$$

$$= \frac{1}{2\Delta}\left\{ \mathrm{shi}\left[\frac{L}{2}\left(i\left(q_k - q_j\right) + \Delta\right)\right] + i\,\mathrm{shi}\left[\frac{L}{2}\left(q_k - q_j - i\Delta\right)\right] - 2i\,\mathrm{shi}\left[\frac{L}{2}\left(q_k - q_j + i\Delta\right)\right] \right\}. \tag{68}$$

The function $K\left(q_k - q_j, L, \Delta\right)$ is an approximation to the Kronecker delta function $\delta_{k,j}$. The function shi is the hyperbolic sine integral $\mathrm{shi}(z) = \int_0^z dt \sinh(t)/t$. A plot of it is shown in **Figure 1**.

Additionally,

$$F(Fh)(v) = \frac{L}{\sqrt{2\Delta}} \sum_{j=-N+1}^{N-1} 2\chi(v,\Delta) e^{iq_j v} \frac{1}{L\sqrt{2\Delta}} \int_{-L/2}^{L/2} e^{-iq_j u} h(u) du$$

$$= \int_{-L/2}^{L/2} du\, h(u) J(v - u, N), \tag{69}$$

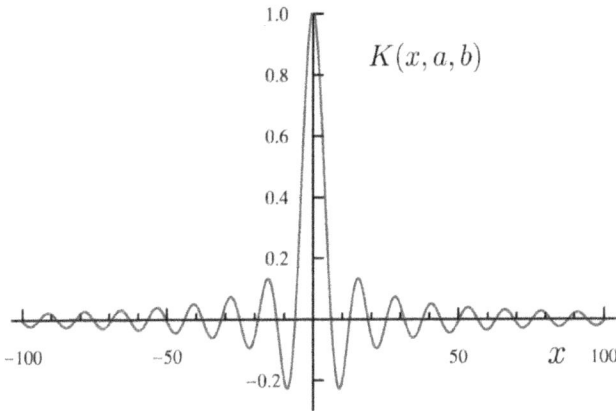

Figure 1. A plot of the kernel function $K(x, a, b)$ with $a = 1$ and $b = .1$. This function is an approximation to the Kronecker delta $\delta_{x,0}$.

where

$$J(x, N) := \frac{2\chi(v, \Delta)}{\Delta} \sum_{j=-N+1}^{N-1} e^{iq_j(v-u)} = \frac{2\chi(v, \Delta)}{\Delta} \sum_{j=-N+1}^{N-1} e^{ij(v-u)\Delta}$$

$$= \frac{2\chi(v, \Delta)}{\Delta} \frac{\sin((N-1/2)(v-u)\Delta)}{\sin((v-u)\Delta/2)},$$

(70)

The ratio of sin functions, in this expression, is an approximation to a series of Dirac delta functions located at $(v-u)\Delta = k\pi$, $k \in \mathbb{N}$. Thus, the operations F and F are finite differences inverse of each other.

6.1. The discrete Fourier transform of the finite differences derivative of a vector

Next, based on Eq. (28), we find that

$$\left(De^{-iqv}g\right)_j = g_{j+1}\left(De^{-iqv}\right)_j + e^{-iq_{j-1}v}(Dg)_j$$

$$= -iv g_{j+1} e^{-iq_j v} + e^{-iq_{j-1}v}(Dg)_j.$$

(71)

If we sum this equality, we get

$$\sum_{j=-N+1}^{N-1} 2\chi(v, \Delta)\left(De^{-iqv}g\right)_j = -iv \sum_{j=-N+1}^{N-1} 2\chi(v, \Delta)g_{j+1}e^{iq_j v}$$

$$+ \sum_{j=-N+1}^{N-1} 2\chi(v, \Delta)e^{-iq_{j-1}v}(Dg)_j$$

(72)

i.e.,

$$(F_N(Dg))(v) = iv(F_{N+1}g)(v)$$

$$+ e^{-iv\Delta} \left[e^{-iq_j v} g_j \Big|_{j=-N+2}^N + \frac{\sqrt{2}}{L} e^{-iq_j v} g_j \Big|_{j=-N+1}^{N-1} \right] \tag{73}$$

Therefore, the discrete Fourier transform of the derivative of a vector **g** is iv times the discrete Fourier transform of **g**, plus boundary terms.

The Fourier transform of the derivative of a continuous function of variable v is easily found if we consider the equality

$$\frac{d}{dv} e^{-iq_j v} = -iq_j e^{-iq_j v}. \tag{74}$$

The integration of this equality with appropriate weights gives

$$-iq_j \int_{-L/2}^{L/2} dv \, e^{-iq_j v} h(v) = - \int_{-L/2}^{L/2} dv \, e^{-iq_j v} \frac{dh(v)}{dv} + e^{-iq_j v} h(v) \Big|_{v=-L/2}^{L/2}, \tag{75}$$

i.e.,

$$(Fh')_j = iq_j (Fh)_j + \frac{1}{L\sqrt{2}} e^{-iq_j v} h(v) \Big|_{v=-L/2}^{L/2}. \tag{76}$$

Hence, as is usual, the Fourier transform of the derivative of a function $h(v)$ of continuous variable v is equal to iq_j times the Fourier transform of the function, plus boundary terms.

7. Conclusion

We proceed with a brief discussion of the relationship between the derivative matrix $\mathbf{D}_N$ and an important concept in quantum mechanics; the concept of *self-adjoint operators* [8, 9]. In particular, we focus on the momentum operator, whose continuous coordinate representation (operation) is given by $-id/dq$, i.e., a derivative times $-i$, in the case of infinite-dimensional Hilbert space.

In the finite-dimensional complex vectorial space (where each vector define a sequence $\{g_i\}_{i=1}^N$ of complex numbers such that $\sum_i |g_i|^2 < \infty$). A transformation $\mathbf{A}$ is usually called *Hermitian*, when its entries $a_{i,j}$ are such that $a_{i,j} = a_{j,i}^*$ (* denote the complex conjugate). Our matrix $\mathbf{D}_N$ is related to an approximation of the derivative (see Section 3) which uses second order finite differences. Therefore, we can ask if the matrix $-i\mathbf{D}_N$ is also Hermitian.

Let $\mathbf{P}_N = -i\mathbf{D}_N$ and $v = ix$ be the eigenvalue of $\mathbf{D}_N$, where $x \in \mathbb{R}$ is a free parameter, the corresponding eigenvalue of $-i\mathbf{D}_N$ is indeed the real value x; which is one of the properties of a Hermitian matrix, as is also the case of infinite-dimensional space (for the Hilbert space on a finite interval, these values are discrete, and for the Hilbert space on the real line, these values conform the continuous spectrum, instead of discrete eigenvalues). Other characteristic of $-i\mathbf{D}_N$ is that the eigenvector corresponding to x is the same exponential function which is the eigenfunction of $-id/dx$ (see Section 2).

Furthermore, let $\mathbf{P}_N^{\dagger}$ denote the adjoint of $\mathbf{P}_N$. Thus, if we restrict our attention to the off-diagonal entries $(\mathbf{P}_N)_{i,j} = -i(\mathbf{D}_N)_{i,j}$, it is fulfilled that $(\mathbf{P}_N^{\dagger})_{i,j} = (-id_{j,i})^* = -id_{i,j} = (\mathbf{P}_N)_{i,j}$ (noticing that, with $v = ix$ then $\chi(x, \Delta) = \sin(x, \Delta)/x \in \mathbb{R}$). Even more, if we do not care about the two entries $d_{i,i}$ for $i = 1, N$, we will have a Hermitian matrix. Finally, as it was seen in Section 4, we can say that $\mathbf{P}_N$ can be considered as a suitable approximation to the conjugate matrix to the coordinate matrix.

In conclusion, we have introduced a matrix with the properties that a Hermitian matrix should comply with, except for two of its entries. Besides, our partition provides congruency between discrete, continuous, and matrix treatments of the exponential function and of its properties.

Author details

Armando Martínez Pérez and Gabino Torres Vega*

*Address all correspondence to: gabino@fis.cinvestav.mx

Physics Department, Cinvestav, México City, México

References

[1] Boole G. A Treatise on the Calculus of Finite Differences. New York: Cambridge University Press; 2009. p. 1860

[2] Harmuth HF, Meffert B. Dogma of the continuum and the calculus of finite differences in quantum physics. In: Advances in Imaging and Electron Physics. Vol. 137. San Diego: Elsevier Academic Press; 2005

[3] Jordan C. Calculus of Finite Differences. 2nd ed. New York: Chelsea Publishing Company; 1950

[4] Richardson CH. An Introduction to the Calculus of Finite Differences. Toronto: D. Van Nostrand; 1954

[5] Santhanam TS, Tekumalla AR. Quantum mechanics in finite dimensions. Foundations of Physics. 1976;**6**:583

[6] Pérez AM, Torres-Vega G. Translations in quantum mechanics revisited. The point spectrum case. Canadian Journal of Physics. 2016;**94**:1365-1368. DOI: 10.1139/cjp-2016-0373

[7] de la Torre AC, Goyeneche D. Quantum mechanics in finite-dimensional Hilbert space. American Journal of Physics. 2003;**71**:49

[8] Gitman DM, Tyutin IV, Voronov BL. Self-Adjoint Extensions in Quantum Mechanics. General Theory and Applications to Schrödinger and Dirac Equations with Singular Potentials. New York: Springer; 2012

[9] Schmüdgen K. Unbounded self-adjoint operators on Hilbert space. In: Graduate Texts in Mathematics. Vol. 265. Heidelberg: Springer; 2012

[10] Hoggatt Jr VE, Bicknell M. Roots of Fibonacci polynomials. Fibonacci Quart. 1973;**11**:271

[11] Li Y. Some properties of Fibonacci and Chebyshev polynomials. Advances in Difference Equations. 2015;**2015**:118

[12] Kaiser G. A friendly guide to wavelets. Birkhäuser. 1994, 2011. ISBN: 978-0-8176-8110-4

Spaces and Linear Systems

Chapter 3

Square Matrices Associated to Mixing Problems ODE Systems

Victor Martinez-Luaces

Additional information is available at the end of the chapter

http://dx.doi.org/10.5772/intechopen.74437

Abstract

In this chapter, mixing problems are considered since they always lead to linear ordinary differential equation (ODE) systems, and the corresponding associated matrices have different structures that deserve to be studied deeply. This structure depends on whether or not there is recirculation of fluids and if the system is open or closed, among other characteristics such as the number of tanks and their internal connections. Several statements about the matrix eigenvalues are analyzed for different structures, and also some questions and conjectures are posed. Finally, qualitative remarks about the differential equation system solutions and their stability or asymptotical stability are included.

Keywords: eigenvalues, Gershgorin circle theorem, mixing problems, linear ODE systems, associated matrices

1. Introduction

Mixing problems (MPs), also known as "compartment analysis" [1], in chemistry involve creating a mixture of two or more substances and then determining some quantity (usually concentration) of the resulting mixture. For instance, a typical mixing problem deals with the amount of salt in a mixing tank. Salt and water enter to the tank at a certain rate, they are mixed with what is already in the tank, and the mixture leaves at a certain rate. This process is modeled by an ordinary differential equation (ODE), as Groestch affirms: "The direct problem for one-compartment mixing models is treated in almost all elementary differential equations texts" [2].

Instead of only one tank, there is a group, as it was stated by Groestch: "The multicompartment model is more challenging and requires the use of techniques of linear algebra" [2].

IntechOpen

© 2018 The Author(s). Licensee IntechOpen. This chapter is distributed under the terms of the Creative Commons Attribution License (http://creativecommons.org/licenses/by/3.0), which permits unrestricted use, distribution, and reproduction in any medium, provided the original work is properly cited. (cc) BY

In particular, the ODE system-associated matrix deserves to be studied since it determines the qualitative behavior of the solutions.

In several previous papers and book chapters [3–6], MPs were studied from different points of view. In the first paper [3], a particular MP with three compartments was proposed, and after applying Laplace transform, this example was connected with important concepts in reactor design, like the transference function. 2 years later, another work [4] analyzed more general MPs in order to obtain characterization results independent of the internal geometry of the tank system. In the third paper [5], the educative potential of MPs was studied, focusing on inverse modeling problems. Finally, in a recent book chapter [6], results for MPs with and without recirculation of fluids were analyzed, and other general results were obtained.

In all these works, a given MP is modeled through an ODE linear system, in which qualitative properties (like stability and asymptotic stability) depend on the eigenvalues and eigenvectors of the associated matrices, so-called MP-matrix.

Taking into account previous results about MP-matrices, and the new ones presented here, two main conjectures can be proposed:

- All the solutions of a given MP are stable.

- If the MP corresponds to an open system, then the solutions are asymptotically stable.

In order to investigate if these conjectures — among others, introduced in the following sections — are true or not, MP-matrices (i.e., square matrices associated to the ODE linear system that models a given MP) should be deeply analyzed.

2. Nomenclature

In this section we introduce a specific terminology useful to allow understanding of the terms properly.

In order to analyze MPs and MP-matrices, we begin by studying a problem already considered in a previous book chapter [6], which involves a tank with five compartments, shown in **Figure 1**.

In this scheme, C_0 is the initial concentration (e.g., salt concentration in water at the entrance of the tank system), C_i is the concentration in the ith compartment ($i = 1,...,5$), and $\Phi_0 \neq 0$ is the incoming and also outgoing flux.

For instance, if Φ_{1k} is the flux that goes from the left (first) to the kth compartment (being $k = 2$, 3, 4) and V_1 is the volume of the first container, then a mass balance gives the following ODE:

$$V_1 \frac{dC_1}{dt} = \Phi_0 C_0 - \Phi_{12}C_1 - \Phi_{13}C_1 - \Phi_{14}C_1 = \Phi_0 C_0 - \left(\Phi_{12} + \Phi_{13} + \Phi_{14}\right)C_1 \tag{1}$$

The ODEs associated with the central compartments ($i = 2, 3, 4$) are simpler, since in each case, there is only one incoming flux Φ_{1k} (being $k = 2, 3, 4$) and a unique outgoing flux Φ_{k5} (being

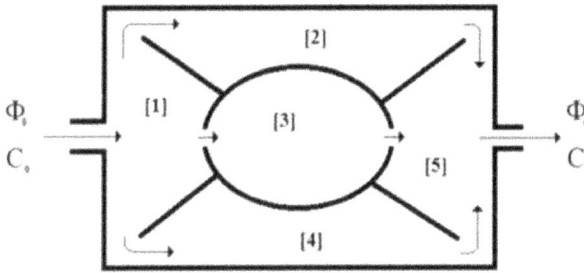

Figure 1. A tank with five internal compartments.

$k = 2, 3, 4$). Once again, if V_k is the volume of the kth container, these equations can be written as.

$$V_2 \frac{dC_2}{dt} = \Phi_{12}C_1 - \Phi_{25}C_2, \ V_3 \frac{dC_3}{dt} = \Phi_{13}C_1 - \Phi_{35}C_3, \ V_4 \frac{dC_4}{dt} = \Phi_{14}C_1 - \Phi_{45}C_4 \qquad (2)$$

Finally, for the right (fifth) container, we have:

$$V_5 \frac{dC_5}{dt} = \Phi_{25}C_2 + \Phi_{35}C_3 + \Phi_{45}C_4 - \Phi_0 C_5 \qquad (3)$$

If all these equations are put together, the following ODE system is obtained:

$$\begin{cases} V_1 \dfrac{dC_1}{dt} = \Phi_0 C_0 - (\Phi_{12} + \Phi_{13} + \Phi_{14})C_1 \\ V_2 \dfrac{dC_2}{dt} = \Phi_{12}C_1 - \Phi_{25}C_2 \\ V_3 \dfrac{dC_3}{dt} = \Phi_{13}C_1 - \Phi_{35}C_3 \\ V_4 \dfrac{dC_4}{dt} = \Phi_{14}C_1 - \Phi_{45}C_4 \\ V_5 \dfrac{dC_5}{dt} = \Phi_{25}C_2 + \Phi_{35}C_3 + \Phi_{45}C_4 - \Phi_0 C_5 \end{cases} \qquad (4)$$

After some algebraic manipulations, the corresponding mathematical model can be written as $\frac{d}{dt}C = AC + C_0 B$, where.

$$C = \begin{pmatrix} C_1 \\ C_2 \\ C_3 \\ C_4 \\ C_5 \end{pmatrix} \text{ and } B = \begin{pmatrix} \Phi_0/V_1 \\ 0 \\ 0 \\ 0 \\ 0 \end{pmatrix} \qquad (5)$$

The system-associated matrix (MP-matrix) is

$$
A = \begin{pmatrix}
-(\Phi_{12} + \Phi_{13} + \Phi_{14})/V_1 & 0 & 0 & 0 & 0 \\
\Phi_{12}/V_2 & -\Phi_{25}/V_2 & 0 & 0 & 0 \\
\Phi_{13}/V_3 & 0 & -\Phi_{35}/V_3 & 0 & 0 \\
\Phi_{14}/V_4 & 0 & 0 & -\Phi_{45}/V_4 & 0 \\
0 & \Phi_{25}/V_5 & \Phi_{35}/V_5 & \Phi_{45}/V_5 & -\Phi_0/V_5
\end{pmatrix}
\tag{6}
$$

Hereafter, we will call MP-matrix to any ODE system-associated matrix related to a given MP, like matrix A of Eq. (6).

In the previous example, the MP-matrix obviously depends on the numbers given to the different containers. In that example it was possible to enumerate the compartments such that the flux always goes from the ith compartment to the jth one, where $i < j$. For instance, a possible enumeration for this purpose is the one illustrated in **Figure 1**.

In general, if in a given MP it is possible to enumerate the containers such that the flux always goes from the ith compartment to the jth one, with $i < j$, then the MP will be considered as a mixing problem without recirculation (MP-WR).

Now, let us analyze a different problem, where a couple of tanks are linked by all possible connections between them, including recirculation from the second tank back to the first one, as in **Figure 2**. This problem represents an interesting variation of an MP analyzed by Zill [7] in his textbook, where the main difference is that this new MP has no incoming and/or outgoing flux, i.e., it is a closed system.

If in a given MP we have that $\sum \Phi_i = 0$, being Φ_i all the system incoming fluxes, and $\sum \Phi_k = 0$, being Φ_k all the system outgoing fluxes, then it will be named MP closed system (MP-CS). Otherwise, it will be an open system (MP-OS).

Taking into account the abovementioned nomenclature, the example considered in **Figure 2** corresponds to an MP-CS, while the MP analyzed in Zill's textbook [7] is an MP-OS, and both are systems with recirculation.

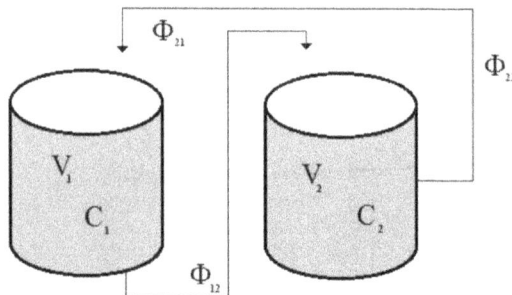

Figure 2. Two tanks with recirculation and no incoming or outgoing fluxes.

Finally, it is important to observe that in both examples (**Figures 1** and **2**), we have $\sum \Phi_i = \sum \Phi_k$, being Φ_i all the system incoming fluxes and Φ_k the corresponding outgoing fluxes. This equation must be satisfied, since the compartments are neither filled up nor emptied with time, at least for the typical MPs' real-life most interesting situations.

In that case all the compartment volumes remain constant, and so if in an MP the following equation $\sum \Phi_i = \sum \Phi_k$ (being Φ_i all the system incoming fluxes and Φ_k the corresponding outgoing fluxes) is satisfied, it will refer to a mixing problem with constant volumes (MP-CV).

Taking into account all these terms, several previous results can be reformulated, as shown in the next section.

3. Previous results revisited

In order to give some general results, it is convenient to consider two different situations: MP without recirculation and MP with recirculation.

Considering again the example in **Figure 1**, it is possible to enumerate the compartments, such that the flux always goes from the ith container to the jth one, being $i < j$, shown in brackets.

Analyzing the system (Eq. (4)), it is easy to observe that for the jth container, the ODE right hand side is a linear combination of a subset of $\{C_0, C_1, \ldots, C_{j-1}, C_j\}$, and this result can be extended straightforward. In fact, in a previous book chapter [6], it was proved that if in a given MP the compartments can be enumerated such that there is no recirculation (i.e., if $i < j$ there is no flux from compartment j to compartment i), then the ODE corresponding to the jth compartment will be of the form:

$$V_j \frac{dC_j}{dt} = \alpha_{i1} C_{i1} + \alpha_{i2} C_{i2} + \ldots + \alpha_{ik} C_{ik} \tag{7}$$

being $\{i1, i2, \ldots, ik\} \subset \{1, 2, \ldots, j\}$ and $\alpha_{i1}, \alpha_{i2}, \ldots, \alpha_{ik} \in R$.

As a consequence, under the previous conditions, the corresponding ODE system has an associated upper matrix.

Revisiting the ODE system (Eq. (4)), corresponding to **Figure 1**, it can be rewritten as

$$
\begin{cases}
\dfrac{dC_1}{dt} = \dfrac{\Phi_0}{V_1} C_0 - \dfrac{(\Phi_{12} + \Phi_{13} + \Phi_{14})}{V_1} C_1 \\[2mm]
\dfrac{dC_2}{dt} = \dfrac{\Phi_{12}}{V_2} C_1 - \dfrac{\Phi_{25}}{V_2} C_2 \\[2mm]
\dfrac{dC_3}{dt} = \dfrac{\Phi_{13}}{V_3} C_1 - \dfrac{\Phi_{35}}{V_3} C_3 \\[2mm]
\dfrac{dC_4}{dt} = \dfrac{\Phi_{14}}{V_4} C_1 - \dfrac{\Phi_{45}}{V_4} C_4 \\[2mm]
\dfrac{dC_5}{dt} = \dfrac{\Phi_{25}}{V_5} C_2 + \dfrac{\Phi_{35}}{V_5} C_3 + \dfrac{\Phi_{45}}{V_5} C_4 - \dfrac{\Phi_0}{V_5} C_5
\end{cases}
\tag{8}
$$

It follows that for the jth compartment, the coefficient corresponding to C_j can be written as $\dfrac{-\sum\limits_{k}\Phi_{jk}}{V_j}$, where $\sum\limits_{k}\Phi_{jk}$ represents the sum of outgoing fluxes. This situation can be easily generalized, since concentration C_j only appears in the right hand side of the corresponding ODE when a certain flux is leaving the tank. Combining this result with the previous one— about the upper matrix—it is easy to observe that the ODE system has only negative eigen-

values of the form $\lambda_j = \dfrac{-\sum\limits_{k}\Phi_{jk}}{V_j} < 0,$ for all $j = 1, 2, \ldots, n$.

However, not all of these results can be extended to MPs with recirculation as will be analyzed in the following subsection.

In previous works [4, 5], a "black box" system was analyzed (see **Figure 3**), in order to obtain a necessary condition to be satisfied by any MP-matrix with any number of compartments and unknown internal geometry. In **Figure 3** Φ_0 and C_0 represent flux and concentration at the input, and Φ_0 is also the output flux (since tanks neither fill up nor empty with time), and C_n is the final concentration. In this system there are n compartments inside the black box with volumes V_i and concentrations C_j, and recirculation fluxes may exist or not.

If all volumes V_i remain constant, by performing a mass balance, it can be proved that.

$$\sum_{i=1}^{n} V_i \frac{dC_i}{dt} = \Phi_0 (C_0 - C_n) \tag{9}$$

Then, Eq. (9) is obtained without any consideration of the internal geometry of the tank system and can be easily verified in the previous example (see **Figure 1**). In fact, by adding the equations of the ODE system (Eq. (4)), it follows straightforward that the condition given in Eq. (9) is satisfied. The same conclusion can be drawn from other possible examples, corresponding to open or closed MPs, with or without recirculation. For instance, in the case schematized in **Figure 2**, the ODE system can be written as follows:

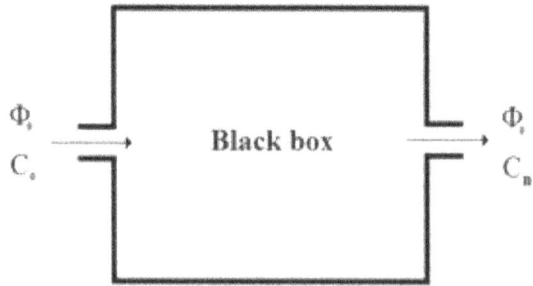

Φ_0
C_0 **Black box** Φ_0
                                       C_n

Figure 3. A "black box" tank system.

$$\begin{cases} \dfrac{dC_1}{dt} = -\dfrac{\Phi_{12}}{V_1} C_1 + \dfrac{\Phi_{21}}{V_1} C_2 \\ \dfrac{dC_2}{dt} = \dfrac{\Phi_{12}}{V_2} C_1 - \dfrac{\Phi_{21}}{V_2} C_2 \end{cases} \tag{10}$$

Operating with these equations, it can be proved that $V_1 \frac{dC_1}{dt} + V_2 \frac{dC_2}{dt} = 0$, which satisfies condition Eq. (9) since Φ_0 is zero.

The previous result can be generalized as follows: in a given MP—with or without recirculation—with input and output concentrations C_0 and C_n, respectively, and being Φ_0 the incoming and outgoing flux, then, independently of the internal geometry, the condition given by Eq. (9) $\sum_{i=1}^{n} V_i \dfrac{dC_i}{dt} = \Phi_0 (C_0 - C_n)$ is satisfied.

An analogous condition may be used to know if a given matrix may or may not be an MP-matrix. For this purpose, let us consider the MP-matrix $\mathbf{A}$, associated to the ODE system given by Eq. (10):

$$\mathbf{A} = \begin{pmatrix} -\dfrac{\Phi_{12}}{V_1} & \dfrac{\Phi_{21}}{V_1} \\ \dfrac{\Phi_{12}}{V_2} & -\dfrac{\Phi_{21}}{V_2} \end{pmatrix} \tag{11}$$

It is easy to observe that

$$(V_1 \ V_2) \begin{pmatrix} -\dfrac{\Phi_{12}}{V_1} & \dfrac{\Phi_{21}}{V_1} \\ \dfrac{\Phi_{12}}{V_2} & -\dfrac{\Phi_{21}}{V_2} \end{pmatrix} = (0 \ \ 0) \tag{12}$$

This equation can be written as $\mathbf{V}^T \mathbf{A} = \mathbf{0}$, being $\mathbf{V}$ the volumes' vector.

If there exists an incoming (and outgoing) flux $\Phi_0 \neq 0$, the last result will change. For instance, if we compute $\mathbf{V}^T \mathbf{A}$, being $\mathbf{V} = (V_1 \ V_2 \ V_3 \ V_4 \ V_5)$ the volumes' vector and $\mathbf{A}$ the MP-matrix corresponding to **Figure 1**, the result will be

$$\mathbf{V}^T \mathbf{A} = (0 \ \ 0 \ \ 0 \ \ 0 \ \ -\Phi_0) \tag{13}$$

It can be noted that Eq. (12) and Eq. (13) are particular cases of the following result: in a given MP—with or without recirculation—with an incoming and outgoing flux $\Phi_0 \neq 0$, the condition $\mathbf{V}^T \mathbf{A} = (0 \ \cdots \ 0 \ -\Phi_0)$ is satisfied, being $\mathbf{V} = (V_1 \ V_2 \ \cdots \ V_n)$ the volumes' vector and $\mathbf{A}$ the MP-matrix.

Then, independently of the internal geometry of the system, the following condition is satisfied:

$$\mathbf{V}^T \mathbf{A} = (0 \ \cdots \ 0 \ -\Phi_0) \tag{14}$$

Now, let us consider again the MP-matrix $\mathbf{A}$, corresponding to the system of **Figure 2**:

$$\mathbf{A} = \begin{pmatrix} -\dfrac{\Phi_{12}}{V_1} & \dfrac{\Phi_{21}}{V_1} \\ \dfrac{\Phi_{12}}{V_2} & -\dfrac{\Phi_{21}}{V_2} \end{pmatrix} \tag{15}$$

If $\mathbf{A}$ is slightly changed only in its first entry, we have the following matrix:

$$\mathbf{A}_\varepsilon = \begin{pmatrix} -\dfrac{\Phi_{12}}{V_1} + \varepsilon & \dfrac{\Phi_{21}}{V_1} \\ \dfrac{\Phi_{12}}{V_2} & -\dfrac{\Phi_{21}}{V_2} \end{pmatrix} \tag{16}$$

It is easy to observe that this new matrix will not satisfy the condition given by Eq. (14). Moreover, there is no MP associated to this matrix $\mathbf{A}_\varepsilon$, since this condition must be satisfied independently of the internal geometry of the system.

As a first consequence, not every square matrix is an MP-matrix. A second observation is that if a given MP-matrix is slightly changed, the result is not necessarily a new MP-matrix.

Furthermore, if volumes V_i and fluxes Φ_i are multiplied by a scale factor, then the MP-matrix $\mathbf{A}$ Eq. (11) remains unchanged, and so, a scale factor in geometry, not in concentrations, produces exactly the same mathematical model.

After interpreting the previous results, we note that when working with MP-matrices, existence, uniqueness, and stability questions for the inverse-modeling problem have negative answers.

The same situation can be observed in many other inverse problems [2], and it is not an exclusive property of compartment analysis.

4. Some considerations about terminology

We start this section explaining three simple and intuitive terms.

Firstly, we will consider that an input tank is a tank with one or more incoming fluxes. Secondly, a tank with one or more outgoing fluxes will be called output tank. Finally, we will say that an internal tank is a tank without incoming and/or outgoing fluxes to or from outside the system.

Taking into account the previous nomenclature, if Φ_{ki} $\forall k = 1, 2, \cdots, m$ represent all the ith tank incoming fluxes, then $\sum \Phi_{ki} \neq 0$ for an input tank, and in the same way, if Φ_{jk} $\forall k = 1, 2, \cdots, m$ represent all the jth tank outgoing fluxes, then $\sum \Phi_{jk} \neq 0$ for an output tank.

Input and output tanks are not mutually exclusive. For instance, in **Figure 4**, the first tank is an input tank, and at same time, it is an output tank, since it has an incoming flux Φ_0 from outside

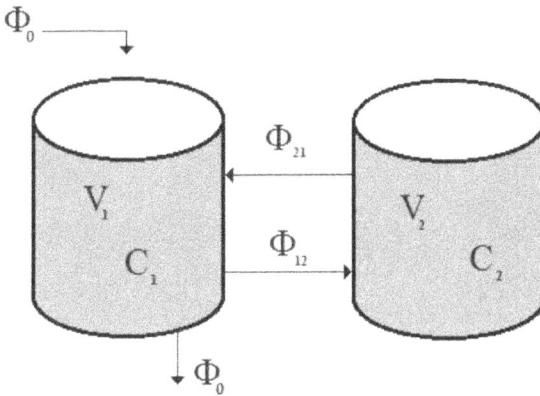

Figure 4. A tank system with recirculation and with incoming and outgoing fluxes.

the system and it also has an outgoing flux Φ_0 that leaves the tank system. It should be noted that in **Figure 4**, the second tank is an internal one.

Another interesting example was proposed by Boelkins et al. [8]. The authors considered a three-tank system connected such that each tank contains an independent inflow that drops salt solution to it, each individual tank has a separated outflow, and each one is connected to the rest of them with inflow and outflow pipes. In this case, all tanks are input and output ones, and there is no internal tank.

It is important to mention that those types of tanks or compartments play different roles in the ODE-associated system and also—as a consequence—in the corresponding MP-matrix. In order to show this fact, let us examine a three-tank system with all the possible connections among them, as in **Figure 5**.

As a first remark, **Figure 5** system has recirculation—unless $\Phi_{21} = \Phi_{32} = \Phi_{31} = 0$, which represents a trivial case—and consequently, an associated upper MP-matrix will not be expected for this problem.

In the mass balance for the first tank—which is an input one—a term $\Phi_0 C_0$ must be considered. In the same way, in the mass balance of the third tank—which is an output one—a term $\Phi_0 C_3$ will appear. These two terms will not be part of the second equation of the ODE system, which can be formulated as follows:

$$\begin{cases} V_1 \dfrac{dC_1}{dt} = \Phi_0 C_0 + \Phi_{21} C_2 + \Phi_{31} C_3 - \left(\Phi_{12} + \Phi_{13}\right) C_1 \\ V_2 \dfrac{dC_2}{dt} = \Phi_{12} C_1 + \Phi_{32} C_3 - \left(\Phi_{21} + \Phi_{23}\right) C_2 \\ V_3 \dfrac{dC_3}{dt} = \Phi_{13} C_1 + \Phi_{23} C_2 - \left(\Phi_{31} + \Phi_{32} + \Phi_0\right) C_3 \end{cases} \qquad (17)$$

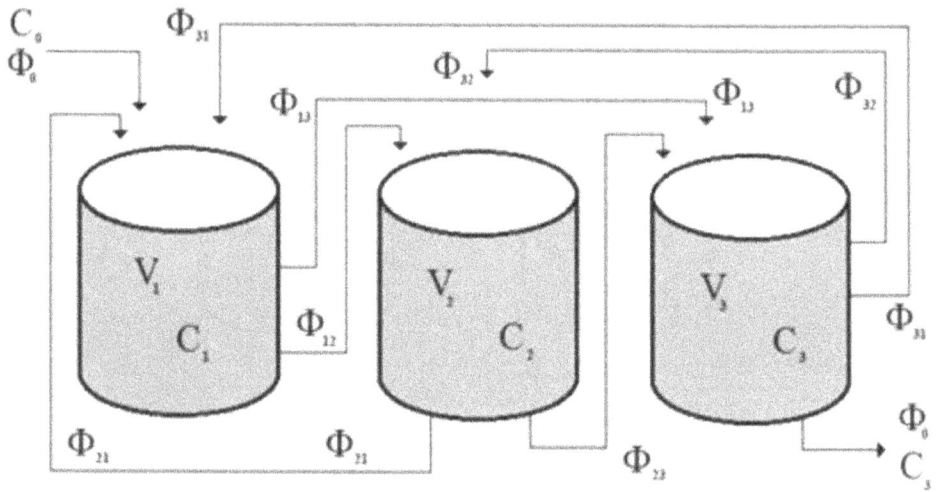

Figure 5. Three tanks with all the possible connections.

Once again, the ODE system can be written as $\dfrac{d}{dt}\mathbf{C} = \mathbf{AC} + C_0\mathbf{B}$, where the MP-matrix is:

$$
\mathbf{A} = \begin{pmatrix}
\left(-1/V_1\right)\left(\Phi_{12}+\Phi_{13}\right) & \Phi_{21}/V_1 & \Phi_{31}/V_1 \\
\Phi_{12}/V_2 & \left(-1/V_2\right)\left(\Phi_{21}+\Phi_{23}\right) & \Phi_{32}/V_2 \\
\Phi_{13}/V_3 & \Phi_{23}/V_3 & \left(-1/V_3\right)\left(\Phi_{31}+\Phi_{32}+\Phi_0\right)
\end{pmatrix}
\tag{18}
$$

In the previous ODE system, the independent vector is:

$$
\mathbf{B} = \begin{pmatrix}
\Phi_0 \Big/ V_1 \\
0 \\
0
\end{pmatrix}
\tag{19}
$$

It is easy to observe that the outgoing flux Φ_0 only appears in the last entry of the MP-matrix $\mathbf{A}$ and the incoming flux Φ_0 only is involved in the first entry of vector $\mathbf{B}$. These facts—particularly the first one—are relevant when applying the Gershgorin circle theorem, which will be exposed in the next section.

5. The Gershgorin circle theorem

The Gershgorin circle theorem first version was published by S. A. Gershgorin in 1931 [9]. This theorem may be used to bind the spectrum of a complex $n \times n$ matrix, and its statement is the following:

Theorem (Gershgorin)

If A is an $n \times n$ matrix, with entries a_{ij} being $i, j \in \{1, \ldots, n\}$, and $R_i = -\sum_{j \neq i} |a_{ij}|$ is the sum of the non-diagonal entry modules in the ith row, then every eigenvalue of A lies within at least one of the closed disks $\overline{D}(a_{ii}, R_i)$, called Gershgorin disks.

This theorem was widely used in previous book chapters [6, 10, 11] in order to obtain new results about matrices corresponding to chemical problems.

Here, the main purpose is to apply this theorem to MP-matrices as a method to bind their eigenvalues, depending on the characteristics of the MP ODE system, and, even more, the compartment considered.

For instance, if we consider the MP corresponding to **Figure 5**, the first ODE of Eq. (17) can be expressed as $\frac{dC_1}{dt} = \frac{\Phi_0}{V_1} C_0 + \frac{\Phi_{21}}{V_1} C_2 + \frac{\Phi_{31}}{V_1} C_3 - \frac{(\Phi_{12} + \Phi_{13})}{V_1} C_1$

This equation—which obviously corresponds to an input tank—gives the first row of the MP-matrix (Eq. (18)) that can be written as $\left(-\dfrac{(\Phi_{12} + \Phi_{13})}{V_1} \quad \dfrac{\Phi_{21}}{V_1} \quad \dfrac{\Phi_{31}}{V_1} \right)$.

The Gershgorin disk corresponding to this row is centered at $a_{11} = -\frac{(\Phi_{12} + \Phi_{13})}{V_1} < 0$ with radius $R_1 = \frac{\Phi_{21} + \Phi_{31}}{V_1}$.

Now, if a flux balance is performed in this input tank, we have this equation: $\Phi_0 + \Phi_{21} + \Phi_{31} = \Phi_{12} + \Phi_{13}$, and then $\Phi_{21} + \Phi_{31} < \Phi_{12} + \Phi_{13}$ (at least if we consider the nontrivial case $\Phi_0 > 0$). As a consequence of this fact, $|a_{11}| > R_1$, and the Gershgorin disk will look like the one schematized in **Figure 6**.

Now, if the second ODE of Eq. (17) is considered, this equation can be written as $\frac{dC_2}{dt} = \frac{\Phi_{12}}{V_2} C_1 + \frac{\Phi_{32}}{V_2} C_3 - \frac{(\Phi_{21} + \Phi_{23})}{V_2} C_2$.

This internal tank equation corresponds to the second row of the MP-matrix (Eq. (18)): $\left(\dfrac{\Phi_{12}}{V_2} \quad -\dfrac{(\Phi_{21} + \Phi_{23})}{V_2} \quad \dfrac{\Phi_{32}}{V_2} \right)$.

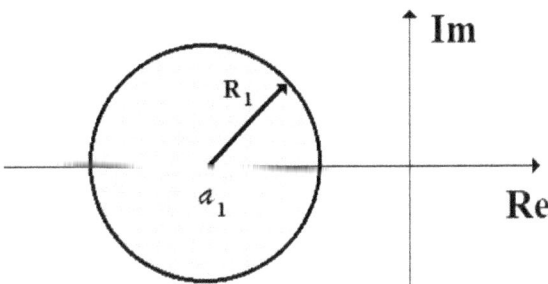

Figure 6. The Gershgorin disk corresponding to an input tank.

The Gershgorin disk corresponding to this row is centered at $a_{22} = -\frac{(\Phi_{21}+\Phi_{23})}{V_2} < 0$ with radius $R_2 = \frac{\Phi_{12}+\Phi_{32}}{V_2}$.

Now, if a flux balance is performed in this internal tank, we have this equation: $\Phi_{12} + \Phi_{32} = \Phi_{21} + \Phi_{23}$, and then $|a_{22}| = R_2$, and the corresponding Gershgorin disk will look like the one schematized in **Figure 7**.

Finally, if the third ODE of Eq. (17) is considered, this equation can be written as $\frac{dC_3}{dt} = \frac{\Phi_{13}}{V_3}C_1 + \frac{\Phi_{23}}{V_3}C_2 - \frac{(\Phi_{31}+\Phi_{32}+\Phi_0)}{V_3}C_3$.

This output tank equation corresponds to the third row of the MP-matrix Eq. (18):
$$\left(\frac{\Phi_{13}}{V_3} \quad \frac{\Phi_{23}}{V_3} \quad -\frac{(\Phi_{31}+\Phi_{32}+\Phi_0)}{V_3} \right).$$

The Gershgorin disk corresponding to this row is centered at the point $a_{33} = -\frac{(\Phi_{31}+\Phi_{32}+\Phi_0)}{V_3} < 0$ with radius $R_3 = \frac{\Phi_{13}+\Phi_{23}}{V_3}$.

The flux balance in this case gives $\Phi_{13} + \Phi_{23} = \Phi_{31} + \Phi_{32} + \Phi_0$, and then $|a_{33}| = R_3$, and the corresponding Gershgorin disk will look like as the one schematized in **Figure 7**.

Taking into account all these results, the Gershgorin circles for the MP of **Figure 5** are shown in **Figure 8**.

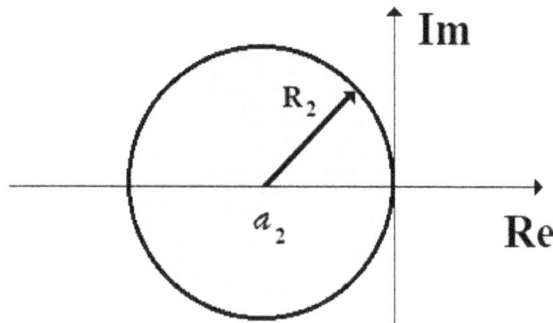

Figure 7. The Gershgorin disk corresponding to an internal tank.

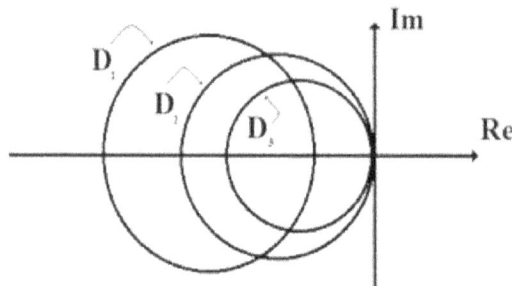

Figure 8. Gershgorin circles for a three-tank system with recirculation.

Since every eigenvalue lies within at least one of the Gershgorin disks, it follows that $\mathrm{Re}(\lambda_i) \le 0$, $\forall i = 1, 2, 3$.

In the following section, these results—among others—will be generalized.

6. The general form of MP-matrices and new results

As stated in Section 3, if there is no recirculation, then the ODE system has only negative eigenvalues of the form $\lambda_j = \dfrac{-\sum\limits_k \Phi_{jk}}{V_j} < 0$, for all $j = 1, 2, \ldots, n$. Then, in this case all the corresponding ODE system solutions will be asymptotically stable.

In a previous work [6], it was proved that in an open MP, with three or less compartments, with or without recirculation, all the corresponding ODE system solutions are asymptotically stable.

It is important to analyze if this result can be generalized or not, when closed systems and/or tanks with more than three compartments are considered. For this purpose, we will start with the following theorem.

Theorem 1

In an open system, if the ith tank is an input one, then the diagonal entry of the ith row is $a_{ii} < 0$ and $|a_{ii}| > R_i$ being $R_i = -\sum\limits_{j \ne i} |a_{ij}|$ the sum of the non-diagonal entry modules of that row.

Proof

If $\Phi_{ai}, \Phi_{bi}, \cdots, \Phi_{ni}$ are the incoming fluxes from other tanks of the system, $\Phi_{iA}, \Phi_{iB}, \cdots, \Phi_{iJ}$ are the outgoing fluxes, and $\Phi_0^1, \Phi_0^2, \cdots, \Phi_0^s$ are the incoming fluxes from outside the system, then the corresponding ODE can be written as

$$V_i \frac{dC_i}{dt} = \Phi_{ai} C_a + \ldots + \Phi_{ni} C_n - (\Phi_{iA} + \ldots + \Phi_{iJ}) C_i + \Phi_0^1 C_0 + \ldots + \Phi_0^s C_s \qquad (20)$$

This equation gives.

$$\frac{dC_i}{dt} = \frac{\Phi_{ai}}{V_i} C_a + \ldots + \frac{\Phi_{ni}}{V_i} C_n - \frac{\sum \Phi_{ij}}{V_i} C_i + \frac{\sum \Phi_0^p}{V_i} C_p \qquad (21)$$

Eq. (20) implies that the ith row of the MP-matrix has entries: $\frac{\Phi_{ki}}{V_i}$ for $k \ne i$, $-\frac{\sum \Phi_{ij}}{V_i}$ for $k = i$, and $\sum \frac{\Phi_0^p}{V_i} C_p$ is part of the independent term.

A flux balance gives $\sum \Phi_{ki} + \sum \Phi_0^p = \sum \Phi_{ij}$, which implies $\sum \Phi_{ki} < \sum \Phi_{ij}$, and then: $a_{ii} = -\frac{\sum \Phi_{ij}}{V_i} < 0$ and also $R_i = \frac{\sum \Phi_{ki}}{V_i} < \frac{\sum \Phi_{ij}}{V_i} = |a_{ii}|$, which proves the theorem.

Corollary 1

In an open system, being the ith tank an input one, the Gershgorin circle corresponding to the ith row looks like the disk in **Figure 6**.

Corollary 2

If in an open system, all are input tanks, all the eigenvalues satisfy the condition $\mathrm{Re}(\lambda_i) < 0$, and the ODE solutions are asymptotically stable.

Theorem 2

In an open system, if the ith tank is not an input one, then the diagonal entry of the ith row is $a_{ii} < 0$ and $|a_{ii}| = R_i$ being $R_i = -\sum_{j \neq i} |a_{ij}|$ the sum of the non-diagonal entry modules of that row.

Proof

If $\Phi_{ai}, \Phi_{bi}, \cdots, \Phi_{ni}$ are the incoming fluxes from other tanks $(a, b, \cdots, n)$ of the MP system, $\Phi_{iA}, \Phi_{iB}, \cdots, \Phi_{iJ}$ are the outgoing fluxes to other tanks $(A, B, \cdots, J)$, and $\Phi_i^1, \Phi_i^2, \cdots, \Phi_i^s$ are the fluxes from the ith tank to outside the system, then the corresponding ODE can be written as

$$V_i \frac{dC_i}{dt} = \Phi_{ai} C_a + \dots + \Phi_{ni} C_n - (\Phi_{iA} + \dots + \Phi_{iJ}) C_i - \Phi_i^1 C_i - \dots - \Phi_i^s C_i \tag{22}$$

This equation gives:

$$\frac{dC_i}{dt} = \frac{\Phi_{ai}}{V_i} C_a + \dots + \frac{\Phi_{ni}}{V_i} C_n - \frac{\sum \Phi_{ij} + \sum \Phi_i^p}{V_i} C_i \tag{23}$$

Eq. (22) implies that the ith row of the MP-matrix has entries $\frac{\Phi_{ki}}{V_i}$ for $k \neq i$ and $-\frac{\sum \Phi_{ij} + \sum \Phi_i^p}{V_i}$ for $k = i$, and this equation does not contribute to the independent term.

In this case a flux balance gives the following equation $\sum \Phi_{ki} = \sum \Phi_{ij} + \sum \Phi_i^p$, then $a_{ii} = -\frac{\sum \Phi_{ij} + \sum \Phi_i^p}{V_i} < 0$, and also $R_i = \frac{\sum \Phi_{ki}}{V_i} = \frac{\sum \Phi_{ij} + \sum \Phi_i^p}{V_i} = |a_{ii}|$, and the theorem is proved.

Corollary 3

In an open system, if the ith tank is not an input one, the Gershgorin circle corresponding to the ith row looks like the disk in **Figure 7**.

Corollary 4

In an open system, the Gershgorin disks look like those of **Figure 8**.

As a consequence of the previous results, the following corollary can be stated.

Corollary 5

In an open system with input and non-input tanks, all the eigenvalues satisfy the condition $\mathrm{Re}(\lambda_i) \leq 0$.

Independently of the previous results, it is easy to observe that all the solutions corresponding to the eigenvalues with $\mathrm{Re}(\lambda_i) < 0$ tend to vanish when $t \to +\infty$.

For this purpose, when analyzing eigenvalues with $\mathrm{Re}(\lambda_i) < 0$, there are two cases to be considered: $\lambda_i \in \mathfrak{R}$ and $\lambda_i \notin \mathfrak{R}$.

In the first case, the corresponding ODE solutions are a linear combination of the functions $\{\exp(-\lambda_i t), \ t\exp(-\lambda_i t), \ t^2\exp(-\lambda_i t), \ldots, \ t^q\exp(-\lambda_i t)\}$, where the number q depends on the algebraic and geometric multiplicity of λ_i (i.e., $AM(\lambda_i)$ and $GM(\lambda_i)$). Taking into account that $\lambda_i < 0$, it follows that $t^n\exp(-\lambda_i t) \underset{t \to +\infty}{\to} 0, \ \forall n = 0, \ 1, \ldots, \ q$.

In the second case—which really happens, as it will be observed later—we have $\lambda_i = a + bi \notin \mathfrak{R}$ (with $a < 0, b \neq 0$). The ODE solutions are a linear combination of $\{\exp(-at)\cos(bt), \ \exp(-at)\sin(bt), \ldots, \ t^q\exp(-at)\cos(bt), \ t^q\exp(-at)\sin(bt)\}$, where the number q depends on $AM(\lambda_i)$ and $GM(\lambda_i)$ as in the other case. It is easy to prove that $t^n\exp(-at)\cos(bt) \underset{t \to +\infty}{\to} 0$ and $t^n\exp(-at)\sin(bt) \underset{t \to +\infty}{\to} 0, \ \forall n = 0, \ 1, \ldots, \ q$, since $a < 0$.

According to the position of the Gershgorin disks for an MP-matrix (see **Figure 8**), the ODE solutions corresponding to an eigenvalue λ_i, with $\mathrm{Re}(\lambda_i) = 0$, can be analyzed.

For this purpose it is important to observe that if an eigenvalue λ_i satisfies $\mathrm{Re}(\lambda_i) = 0$, then it must be $\lambda_i = 0$, since the Gershgorin disks look like those in **Figure 8**.

In this case the ODE solutions are a linear combination of the following functions: $\{\exp(-0t), \ t\exp(-0t), \ t^2\exp(-0t), \ldots, \ t^q\exp(-0t)\} = \{1, \ t, \ t^2, \ldots, \ t^q\}$, where the number q depends on $AM(0)$ and $GM(0)$. In other words, the corresponding solutions are polynomial, and so, they will not tend to vanish nor remain bounded when $t \to +\infty$, unless $AM(0) = MG(0)$, and the polynomial becomes a constant.

Considering all these results, it is obvious that the stability of the ODE system solutions will depend exclusively on $AM(0)$ and $GM(0)$.

7. Several questions and a conjecture

In the previous section, some particular cases with $\lambda_i = 0$ and/or $\lambda_i = a + bi \notin \mathfrak{R}$ (with $a < 0, b \neq 0$) were considered. A first question to analyze is if there exists an MP that satisfies any of these conditions. For this purpose, let us consider the closed MP of **Figure 9**, in which ODE system can be written as $\dfrac{d}{dt}C = AC$, and the corresponding MP-matrix is

$$\begin{pmatrix} -a & 0 & a \\ b & -b & 0 \\ 0 & c & -c \end{pmatrix}, \text{ being } a = \tfrac{\Phi}{V_1}, \ b = \tfrac{\Phi}{V_2}, \text{ and } c = \tfrac{\Phi}{V_3}. \text{ If } \Phi \text{ and } V_i \text{ are chosen such that } a = 1,$$

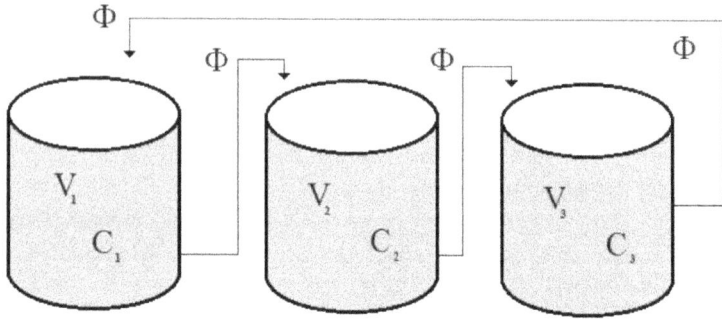

Figure 9. Three tanks with all the possible connections.

$b = 2$, and $c = 3$, it is easy to show that the eigenvalues are $\lambda_1 = 0$ and $\lambda_{2,3} = -3 \pm i\sqrt{2}$, which prove that null and/or complex eigenvalues are possible.

Other questions are not so simple like the previous one. The next two examples propose challenging problems that deserve to be studied:

Question 1:

Is it possible to find an MP-matrix with an eigenvalue $\lambda_i = 0$ such that $AM(0) > 1$?

Question 2:

Is it possible to find an MP-matrix such that $AM(0) > GM(0)$?

Question 3:

Is it possible to find an MP-matrix with complex eigenvalues in an open system?

Finally, it is interesting to observe that all cases analyzed here with $\lambda_i = 0$ correspond to closed systems. Moreover, in a previous book chapter [6], it was proved that $\mathrm{Re}(\lambda_i) \leq 0$, $\forall i$, in any MP open system with three tanks or less. Taking into account all these facts, it can be conjectured that in an open system, all the MP-matrix eigenvalues have negative real part and as a consequence, all the solutions are asymptotically stable.

8. Conclusions

Mixing problems are interesting sources for applied research in mathematical modeling, ODE, and linear algebra, and—as it was shown—their behavior depends on how they are connected. It has been proved that null eigenvalues are not expected in open systems with three or less components, and $\mathrm{Re}(\lambda_i) \leq 0$, $\forall i$ is a general conclusion for open MP-matrices that can be obtained by applying the Gershgorin circle theorem.

As a final remark, all the MP differential equation systems considered in this chapter have stable or asymptotically stable solutions. Nevertheless, this situation may change depending on the answers to the questions and the conjecture presented in the last section, giving a challenging proposal for further research on this topic.

Acknowledgements

The author wishes to thank Marjorie Chaves for her assistance and support in this work.

Author details

Victor Martinez-Luaces

Address all correspondence to: victoreml@gmail.com

Faculty of Engineering, UdelaR, Montevideo, Uruguay

References

[1] Braun M. Differential Equations and their Applications. 3rd ed. New York: Springer; 2013. 546 p. DOI: 10.1007/978-1-4684-9229-3

[2] Groestch C. Inverse Problems: Activities for Undergraduates. Washington D.C.: Mathematical Association of America; 1999. 222 p

[3] Martinez-Luaces V. Engaging secondary school and university teachers in modelling: Some experiences in South American countries. International Journal of Mathematical Education in Science and Technology. 2005;**36**(2–3):193-205

[4] Martinez-Luaces V. Inverse-modelling problems in chemical engineering courses. In: Proceedings of the Southern Hemisphere Conference on Undergraduate Mathematics and Statistics Teaching and Learning. (Delta '07); 26–30 November 2007; Calafate: Argentina. pp. 111-118

[5] Martinez-Luaces V. Modelling and inverse-modelling: Experiences with O.D.E. Linear systems in engineering courses. International Journal of Mathematical Education in Science and Technology. 2009;**40**(2):259-268

[6] Martinez-Luaces V. Matrices in chemical problems: Characterization, properties and consequences about the stability of ODE Systems. In: Baswell A, editor. Advances in Mathematics Research. New York: Nova Publishers; 2017. pp. 1-33

[7] Zill D. Differential Equations with Boundary-Value Problems. 9th ed. Boston: Cengage Learning; 2016. 630 p

[8] Boelkins M, Goldberg J, Potter M. Differential Equations with Linear Algebra. New York: Oxford University Press; 2009. 576 p

[9] Varga R. Geršgorin and His Circles. Berlin, Germany: Springer-Verlag; 2004

[10] Martinez-Luaces V. First order chemical kinetics matrices and stability of O.D.E. systems. In: Kyrchei I, editor. Advances in Linear Algebra Research. New York: Nova Publishers; 2015. pp. 325-343

[11] Martinez-Luaces V. Qualitative behavior of concentration curves in first order chemical kinetics mechanisms. In: Taylor J, editor. Advances in Chemistry Research. New York: Nova Publishers; 2017. pp. 139-169

Chapter 4

Nullspace of Compound Magic Squares

Saleem Al-Ashhab

Additional information is available at the end of the chapter

http://dx.doi.org/10.5772/intechopen.74678

Abstract

In this chapter, we consider special compound $4n \times 4n$ magic squares. We determine a $2n - 3$ dimensional subspace of the nullspace of the $4n \times 4n$ squares. All vectors in the subspaces possess the property that the sum of all entries of each vector equals zero.

Keywords: null space, magic squares, mathematical induction

1. Introduction

A semi-magic square is an $n \times n$ matrix such that the sum of the entries in each row and column is the same. The common value is called the magic constant. If, in addition, the sum of all entries in each left-broken diagonal and each right-broken diagonal is the magic constant, then we call the matrix a pandiagonal magic square. Rosser and Walker show that a pandiagonal 4×4 magic square with magic constant $2s$ has in general the following structure.

A	B	C	ω
E	θ	ς	ρ
$s - C$	$s - \omega$	$s - A$	$s - B$
$s - \varsigma$	$s - \rho$	$s - E$	$s - \theta$

where

$$\omega = 2s - A - B - C;$$

$$\theta = 2s - A - B - E;$$

© 2018 The Author(s). Licensee IntechOpen. This chapter is distributed under the terms of the Creative Commons Attribution License (http://creativecommons.org/licenses/by/3.0), which permits unrestricted use, distribution, and reproduction in any medium, provided the original work is properly cited. (cc) BY

IntechOpen

$$\varsigma = A + E - C;$$

$$\rho = B + C - E.$$

This result was developed by Rosser and Walker. Hendricks proved that the determinant of a pandiagonal magic square is zero. We note that every antipodal pair of elements add up to one-half of the magic constant. Al-Amerie considered in his M.Sc thesis some of the results here. There are three fundamental primitive pandiagonal squares which are 4×4. Kraitchik (see [3, 8]) has shown how to derive all pandiagonal squares from three particular ones.

We define a certain class of 6×6 magic squares, which has a similar structure to the structure of a pandiagonal 4×4 magic square. In this class each antipodal pair will add up to one-third of the magic constant. Precisely, we have:

Definition 1: A 6×6 magic square with $3s$ as a magic constant is called panmagic if

$$a_{ij} + a_{kl} = s, \text{ for each } i, j, k, l \text{ such that } i \equiv k \ (\text{mod } 3) \text{ and } j \equiv l \ (\text{mod } 3).$$

The following matrix is a possible form for this kind of squares:

M	R	W	T	L	K
Q	J	I	H	G	F
P	E	D	C	B	A
$s - T$	$s - L$	$s - K$	$s - M$	$s - R$	$s - W$
$s - H$	$s - G$	$s - F$	$s - Q$	$s - J$	$s - I$
$s - C$	$s - B$	$s - A$	$s - P$	$s - E$	$s - D$

where

$$M = J + I + H + E + D + C - L - K - \frac{3s}{2},$$

$$W = K - I + F - D + A,$$

$$P = 3s - E - D - C - B - A$$

$$Q = 3s - J - I - H - G - F,$$

$$R = L - J + G - E + B,$$

$$T = \frac{9s}{2} - L - K - H - G - F - C - B - A.$$

Note that we have the following relations:

$$\begin{aligned} M + Q + P &= T + H + C, \\ R + J + E &= L + G + B, \\ W + I + D &= K + F + A. \end{aligned} \tag{1}$$

Using Maple we can show that the 6×6 panmagic square possesses a nontrivial null space, which can be written in the following form:

$$\left\{ z(x_1, x_2, x_3, -x_1, -x_2, -x_3)' : z \in R \right\}$$

where

$$x_1 = (A - D)(G - J)(B - E)(I - F),$$

$$x_2 = (F - I)(B + 2C + E - 3s) + 2FD - 2AI + (D - A)(G + J + 2H - 3s),$$

$$x_3 = (B - E)(F + I + 2H) + (A + D + 2C + 2B + 2E - 3s)(J - G).$$

Note that the sum of all entries of the vectors is zero. For example:

-51	39	26	0	9	13
54	-10	-2	-5	4	-5
-5	1	2	3	17	18
12	3	-1	63	-27	-14
17	8	17	-42	22	14
9	-5	-6	17	11	10

has as nullspace $\left\{ z(34, 115, -132, -34, -115, 132)^t : z \in R \right\}$.

Definition 2: A 8×8 square consisting of 4 pandiagonal magic squares $A_{11}, A_{12}, A_{21}, A_{22}$ having the same magic sum in the form

$$\begin{bmatrix} A_{11} & A_{12} \\ A_{21} & A_{22} \end{bmatrix}$$

is called a compound magic square if the following relation holds:

$$A_{22} + A_{11} = A_{12} + A_{21}.$$

It is easy to check if the last relation guarantees that the square is a magic 8×8 square. In the same manner we can combine four panmagic squares in a magic square.

Definition 3: Let $B_{22}, B_{11}, B_{12}, B_{21}$ be panmagic squares having the same magic constant. Assume that $B_{22} + B_{11} = B_{12} + B_{21}$. Then the matrix

$$\begin{bmatrix} B_{11} & B_{12} \\ B_{21} & B_{22} \end{bmatrix}$$

is called the compound 12×12 magic square.

The condition $B_{22} + B_{11} = B_{12} + B_{21}$ ensures that the compound 12×12 magic square is magic.

2. Main results

We prove first a simple result for a compound square of 4×4 squares. We then generalize this result for an arbitrary number of squares.

Proposition 1: The compound 8×8 magic square processes a three-dimensional subspace of its nullspace.

Proof: First we note that the vector

$$(1, 1, 1, 1, -1, -1, -1, -1)'$$

is a nonzero vector, which belongs to the nullspace of the square, since the squares have the same magic constant.

Now, the square A_{11} (res. A_{12}) has a nonzero vector v_{11} (res. v_{12}), which belongs to the nullspace of the square, since A_{11}(res. A_{12}) is a pandiagonal magic square. We look for four numbers $f_{11}, f_{12}, f_{21}, f_{22}$ such that the vector

$$\begin{pmatrix} f_{11}v_{11} + f_{12}v_{12} \\ f_{21}v_{11} + f_{22}v_{12} \end{pmatrix}$$

belongs to the nullspace of the square. To do this we compute the following matrix multiplication:

$$\begin{bmatrix} A_{11} & A_{12} \\ A_{21} & A_{22} \end{bmatrix} \begin{pmatrix} f_{11}v_{11} + f_{12}v_{12} \\ f_{21}v_{11} + f_{22}v_{12} \end{pmatrix} = \begin{pmatrix} A_{11}(f_{11}v_{11} + f_{12}v_{12}) + A_{12}(f_{21}v_{11} + f_{22}v_{12}) \\ A_{21}(f_{11}v_{11} + f_{12}v_{12}) + A_{22}(f_{21}v_{11} + f_{22}v_{12}) \end{pmatrix}$$

According to the choice of v_{11} and v_{12} we obtain the vector $(g_1, g_2)'$ as the result of matrix multiplication, where:

$$g_1 = A_{11}f_{12}v_{12} + A_{12}f_{21}v_{11},$$
$$g_2 = A_{21}v_{11}f_{11} + A_{21}v_{12}f_{12} + (A_{12}v_{11} + A_{21}v_{11})f_{21} + (A_{21}v_{12} - A_{11}v_{12})f_{22}.$$

Note that we used the relation $A_{22} = A_{12} + A_{21} - A_{11}$. We can rewrite the vector $(g_1, g_2)'$ in the form.

$$\begin{bmatrix} 0 & A_{11}v_{12} & A_{12}v_{11} & 0 \\ A_{21}v_{11} & A_{21}v_{12} & (A_{12} + A_{21})v_{11} & (A_{21} - A_{11})v_{12} \end{bmatrix} \begin{pmatrix} f_{11} \\ f_{12} \\ f_{21} \\ f_{22} \end{pmatrix} \tag{2}$$

According to Al-Ashhab (see [3]) we can assume that the vectors in the nullspace of the pandiagonal magic square are

$$v_{ij} = \left(v_{ij}^*, v_{ij}^{**}, -v_{ij}^*, -v_{ij}^{**}\right)', \text{ for } i = 1, j = 1, 2$$

Further, we can assume that

$$A_{ij} = \begin{bmatrix} a_{ij} & b_{ij} & c_{ij} & d_{ij} \\ e_{ij} & f_{ij} & g_{ij} & h_{ij} \\ s - c_{ij} & s - d_{ij} & s - a_{ij} & s - b_{ij} \\ s - g_{ij} & s - h_{ij} & s - e_{ij} & s - f_{ij} \end{bmatrix}, i, j = 1, 2$$

Hence, we can assume that:

$$A_{ij}v_{ij} = \begin{pmatrix} a_{ij}v_{ij}^* + b_{ij}v_{ij}^{**} - c_{ij}v_{ij}^* - d_{ij}v_{ij}^{**} \\ e_{ij}v_{ij}^* + f_{ij}v_{ij}^{**} - g_{ij}v_{ij}^* - h_{ij}v_{ij}^{**} \\ -c_{ij}v_{ij}^* - d_{ij}v_{ij}^{**} + a_{ij}v_{ij}^* + b_{ij}v_{ij}^{**} \\ -g_{ij}v_{ij}^* - h_{ij}v_{ij}^{**} + e_{ij}v_{ij}^* + f_{ij}v_{ij}^{**} \end{pmatrix} = \begin{pmatrix} \left(a_{ij} - c_{ij}\right)v_{ij}^* + \left(b_{ij} - d_{ij}\right)v_{ij}^{**} \\ \left(e_{ij} - g_{ij}\right)v_{ij}^* + \left(f_{ij} - h_{ij}\right)v_{ij}^{**} \\ -\left(a_{ij} - c_{ij}\right)v_{ij}^* - \left(b_{ij} - d_{ij}\right)v_{ij}^{**} \\ -\left(e_{ij} - g_{ij}\right)v_{ij}^* - \left(f_{ij} - h_{ij}\right)v_{ij}^{**} \end{pmatrix}$$

Since the sum of two pandiagonal magic squares is pandiagonal magic, we deduce that four rows in the matrix in Eq. (2) are redundant. Since we have the relations

$$a_{11} + e_{11} = c_{11} + g_{11} \Rightarrow a_{11} - c_{11} = -\left(e_{11} - g_{11}\right)$$
$$b_{11} + f_{11} = d_{11} + h_{11} \Rightarrow b_{11} - d_{11} = -\left(f_{11} - h_{11}\right)$$

the application of elementary row operations on the matrix in Eq. (2) yields to

$$\begin{bmatrix} 0 & r_{12} & r_{21} & 0 \\ q_{11} & q_{12} & r_{21} + q_{11} & q_{12} - r_{12} \\ 0 & 0 & 0 & 0 \\ 0 & 0 & 0 & 0 \\ 0 & 0 & 0 & 0 \\ 0 & 0 & 0 & 0 \\ 0 & 0 & 0 & 0 \\ 0 & 0 & 0 & 0 \end{bmatrix}$$

where

$$r_{12} = \left(a_{11} - c_{11}\right)v_{12}^* + \left(b_{11} - d_{11}\right)v_{12}^{**}$$
$$r_{21} = \left(a_{12} - c_{12}\right)v_{11}^* + \left(b_{12} - d_{12}\right)v_{11}^{**}$$
$$q_{11} = \left(a_{21} - c_{21}\right)v_{11}^* + \left(b_{21} - d_{21}\right)v_{11}^{**}$$
$$q_{12} = \left(a_{21} - c_{21}\right)v_{12}^* + \left(b_{21} - d_{21}\right)v_{12}^{**}$$

This analysis enables us to conclude the following relations from (2):

$$f_{11} = -\frac{r_{12}r_{21} + q_{11}r_{12} - q_{12}r_{21}}{q_{11}r_{12}}f_{21} + \frac{-q_{12} + r_{12}}{q_{11}}f_{22}, f_{12} = -\frac{r_{21}}{r_{12}}f_{21}.$$

If we set

$$f_{12} = 0, \; f_{21} = 0, \; f_{22} = q_{11}, \; f_{11} = r_{12} - q_{12},$$

which is consistent with the previous relations, we conclude that the vector

$$\begin{pmatrix} (r_{12} - q_{12})v_{11} \\ q_{11}v_{12} \end{pmatrix}$$

belongs to the nullspace of the square. We can make another choice as follows.

$$f_{22} = 0, f_{21} = r_{12}q_{11}, f_{12} = -r_{21}q_{11}, f_{11} = r_{21}q_{12} - r_{12}(r_{21} + q_{11})$$

and we obtain a vector belonging to the nullspace of the square, which is

$$\begin{pmatrix} (r_{21}q_{12} - r_{12}(r_{21} + q_{11}))v_{11} - r_{21}q_{11}v_{12} \\ -r_{21}q_{11}v_{11} \end{pmatrix}$$

Now, the vectors v_{12}, v_{11} are linearly independent, since they correspond to different magic squares. Hence, the last two vectors are linearly independent. Also the vector

$$(1, 1, 1, 1, -1, -1, -1, -1)'$$

is linearly independent with the last two vectors, since its first two entries are not the opposite of the third and fourth entry. □

For example, the following square is a compound 8×8 magic square.

0	14	−19	13	10	5	−22	15
−12	6	7	7	−20	13	12	3
23	−9	4	−10	26	−11	−6	−1
−3	−3	16	−2	−8	1	24	−9
−16	25	−17	16	−6	16	−20	18
1	−2	2	7	−7	5	7	3
21	−12	20	−21	24	−14	10	−12
2	−3	3	6	−3	1	11	−1

For this square we can construct as described the following two vectors in its nullspace

$$\left(-\frac{38}{5}, \frac{722}{5}, \frac{38}{5}, -\frac{722}{5}, -170, -544, 170, 544\right)^{t}$$

$$\left(-\frac{216006}{5}, -\frac{85026}{5}, \frac{216006}{5}, \frac{85026}{5}, 3774, -71706, -3774, 71706\right)^{t}$$

In fact, its nullity is 3. Thus, these two vectors together with

$$(1, 1, 1, 1, -1, -1, -1, -1)'$$

form a basis of its nullspace.

We prove now a similar result to the previous proposition, where we replace the 4×4 square with a 6×6 one.

Proposition 2: The compound 12×12 magic square possess a three-dimensional subspace of its nullspace.

Proof: First we note that the vector

$$(1, 1, 1, 1, 1, 1 - 1, -1, -1, -1, -1, -1)'$$

is a nonzero vector, which belongs to the nullspace of the square, since the squares have the same magic constant.

We look for scalars $v_1, v_2, v_3, v_4, v_5, v_6$ such that

$$
\begin{bmatrix}
a_{11} & b_{11} & c_{11} & d_{11} & e_{11} & f_{11} & a_{12} & b_{12} & c_{12} & d_{12} & e_{12} & f_{12} \\
g_{11} & h_{11} & i_{11} & j_{11} & k_{11} & l_{11} & g_{12} & h_{12} & i_{12} & j_{12} & k_{12} & l_{12} \\
m_{11} & n_{11} & o_{11} & p_{11} & q_{11} & r_{11} & m_{12} & n_{12} & o_{12} & p_{12} & q_{12} & r_{12} \\
s-d_{11}s-e_{11}s-f_{11}s-a_{11}s-b_{11}s-c_{11} & s-d_{12}s-e_{12}s-f_{12}s-a_{12}s-b_{12}s-c_{12} \\
s-j_{11}s-k_{11}s-l_{11}s-g_{11}s-h_{11}s-i_{11} & s-j_{12}s-k_{12}s-l_{12}s-g_{12}s-h_{12}s-i_{12} \\
s-p_{11}s-q_{11}s-r_{11}s-m_{11}s-n_{11}s-o_{11} & s-p_{12}s-q_{12}s-r_{12}s-m_{12}s-n_{12}s-o_{12} \\
a_{21} & b_{21} & c_{21} & d_{21} & e_{21} & f_{21} & a_{22} & b_{22} & c_{22} & d_{22} & e_{22} & f_{22} \\
g_{21} & h_{21} & i_{21} & j_{21} & k_{21} & l_{21} & g_{22} & h_{22} & i_{22} & j_{22} & k_{22} & l_{22} \\
m_{21} & n_{21} & o_{21} & p_{21} & q_{21} & r_{21} & m_{22} & n_{22} & o_{22} & p_{22} & q_{22} & r_{22} \\
s-d_{21}s-e_{21}s-f_{21}s-a_{21}s-b_{11}s-c_{21} & s-d_{22}s-e_{22}s-f_{22}s-a_{22}s-b_{22}s-c_{22} \\
s-j_{21}s-k_{21}s-l_{21}s-g_{21}s-h_{21}s-i_{21} & s-j_{22}s-k_{22}s-l_{22}s-g_{22}s-h_{22}s-i_{22} \\
s-p_{21}s-q_{21}s-r_{21}s-m_{21}s-n_{21}s-o_{21} & s-p_{22}s-q_{22}s-r_{22}s-m_{22}s-n_{22}s-o_{22}
\end{bmatrix}
\cdot
\begin{bmatrix}
v_1 \\ v_2 \\ v_3 \\ -v_1 \\ -v_2 \\ -v_3 \\ v_4 \\ v_5 \\ v_6 \\ -v_4 \\ -v_5 \\ -v_6
\end{bmatrix}
=
\begin{bmatrix}
0 \\ 0 \\ 0 \\ 0 \\ 0 \\ 0 \\ 0 \\ 0 \\ 0 \\ 0 \\ 0 \\ 0
\end{bmatrix}
$$

We transform this equation into a linear system, in which we eliminate the redundant equations. The system becomes

$$(a_{11} - d_{11})v_1 + (h_{11} - e_{11})v_2 + (c_{11} - f_{11})v_3 + (a_{12} - d_{12})v_4 + (b_{12} - e_{12})v_5 + (c_{12} - f_{12})v_6 = 0$$
$$(g_{11} - j_{11})v_1 + (h_{11} - k_{11})v_2 + (i_{11} - l_{11})v_3 + (g_{12} - j_{12})v_4 + (h_{12} - k_{12})v_5 + (l_{12} - l_{12})v_6 = 0$$
$$(m_{11} - p_{11})v_1 + (n_{11} - q_{11})v_2 + (o_{11} - r_{11})v_3 + (m_{12} - p_{12})v_4 + (n_{12} - q_{12})v_5 + (o_{12} - r_{12})v_6 = 0$$
$$(a_{21} - d_{21})v_1 + (b_{21} - e_{21})v_2 + (c_{21} - f_{21})v_3 + (a_{22} - d_{22})v_4 + (b_{22} - e_{22})v_5 + (c_{22} - f_{22})v_6 = 0$$
$$(g_{21} - j_{21})v_1 + (h_{21} - k_{21})v_2 + (i_{21} - l_{21})v_3 + (g_{22} - j_{22})v_4 + (h_{22} - k_{22})v_5 + (i_{22} - l_{22})v_6 = 0$$
$$(m_{21} - p_{21})v_1 + (n_{21} - q_{21})v_2 + (o_{21} - r_{21})v_3 + (m_{22} - p_{22})v_4 + (n_{22} - q_{22})v_5 + (o_{22} - r_{22})v_6 = 0$$

From the definition of the panmagic square we know that

$$a_{ij} + g_{ij} + m_{ij} = d_{ij} + j_{ij} + p_{ij} \implies \left(a_{ij} - d_{ij}\right) + \left(g_{ij} - j_{ij}\right) = -\left(m_{ij} - p_{ij}\right) \tag{3}$$

$$b_{ij} + h_{ij} + n_{ij} = e_{ij} + k_{ij} + q_{ij} \implies \left(b_{ij} - e_{ij}\right) + \left(h_{ij} - e_{ij}\right) = -\left(n_{ij} - q_{ij}\right) \tag{4}$$

$$c_{ij} + l_{ij} + o_{ij} = f_{ij} + l_{ij} + r_{ij} \implies \left(c_{ij} - f_{ij}\right) + \left(i_{ij} - l_{ij}\right) = -\left(o_{ij} - r_{ij}\right) \tag{5}$$

Thus, due to Eqs. (3)–(5), we can reduce the linear system to the following

$$(a_{11} - d_{11})v_1 + (b_{11} - e_{11})v_2 + (c_{11} - f_{11})v_3 + (a_{12} - d_{12})v_4 + (b_{12} - e_{12})v_5 + (c_{12} - f_{12})v_6 = 0$$
$$(g_{11} - j_{11})v_1 + (h_{11} - k_{11})v_2 + (i_{11} - l_{11})v_3 + (g_{12} - j_{12})v_4 + (h_{12} - k_{12})v_5 + (i_{12} - l_{12})v_6 = 0$$
$$(a_{21} - d_{21})v_1 + (b_{21} - e_{21})v_2 + (c_{21} - f_{21})v_3 + (a_{22} - d_{22})v_4 + (b_{22} - e_{22})v_5 + (c_{22} - f_{22})v_6 = 0$$
$$(g_{21} - j_{21})v_1 + (h_{21} - k_{21})v_2 + (i_{21} - l_{21})v_3 + (g_{22} - j_{22})v_4 + (h_{22} - k_{22})v_5 + (i_{22} - l_{22})v_6 = 0$$

We can verify using the computer that the coefficient matrix of this system has in general the rank four. Hence, we deduce that v_1, v_2, v_3, v_4 depends on v_5 and v_6. By letting v_5 and v_6 take the values 0 and 1 we obtain two linearly independent vectors in the nullspace. These two vectors do not possess the property that the first six elements are the opposite of the last six elements. Hence, they are independent of the vector $(1,1,1,1,1,1-1,-1,-1,-1,-1,-1)'$. $\square$

Remark: We did not here make use of the relation $B_{22} + B_{11} = B_{12} + B_{21}$. It actually does not affect the proof.

For example, the following square is a compound 12×12 magic square.

−51	39	26	0	9	13	6	17	15	−6	0	4
54	−10	−2	−5	4	−5	20	5	2	0	9	0
−5	1	2	3	17	18	−24	6	7	8	19	20
12	3	−1	63	−27	−14	18	12	8	6	−5	−3
17	8	17	−42	22	14	12	3	12	−8	7	10
9	−5	−6	17	11	10	4	−7	−8	36	6	5
2	53	45	−131	33	34	59	31	34	−137	24	25
−10	0	10	11	12	13	−44	15	14	16	17	18
−89	21	22	23	29	30	−108	26	27	28	31	32
143	−21	−22	10	−41	−33	149	−12	−13	−47	−19	−22
1	0	−1	22	12	2	−4	−5	−6	56	−3	−2
−11	−17	−18	101	−9	−10	−16	−19	−20	120	−14	−15

Using the computer we can verify that its nullity is 3. In other words, the constructed subspace is the nullspace itself.

We can generalize the previous result for an arbitrary number of squares involved in the compound square.

Theorem 1: Let A_{ij} be the distinct pandiagonal magic square with magic constant $2s$ having the structure:

$$A_{ij} = \begin{bmatrix} a_{ij} & b_{ij} & c_{ij} & d_{ij} \\ e_{ij} & f_{ij} & g_{ij} & h_{ij} \\ s - c_{ij} & s - d_{ij} & s - a_{ij} & s - b_{ij} \\ s - g_{ij} & s - h_{ij} & s - e_{ij} & s - f_{ij} \end{bmatrix}$$

such that $A_{ij} = A_{1j} + A_{i1} - A_{11}$ for $i, j = 1, ..., n$. Assume that $(a_{11} + c_{12} - c_{11} - a_{12}) \neq 0$. Then, the following $4n \times 4n$ matrix

$$\begin{bmatrix} A_{11} A_{12} & A_{13}...A_{1n} \\ A_{21} A_{22} & A_{23}...A_{2n} \\ A_{31} A_{32} & A_{33}...A_{3n} \\ \vdots & \vdots & \vdots & \vdots & \vdots \\ A_{n1} A_{n2} & A_{n3}...A_{nn} \end{bmatrix}$$

possesses a $2n - 3$ dimensional subspace of its nullspace, which is generated by the vectors

$$\begin{pmatrix} b_{11}\text{-}d_{11}\text{-}b_{12} + d_{12} \\ -(a_{11} + c_{12} - c_{11} - a_{12}) \\ -(b_{11}\text{-}d_{11}\text{-}b_{12} + d_{12}) \\ (a_{11} + c_{12} - c_{11} - a_{12}) \\ -(b_{11}\text{-}d_{11}\text{-}b_{12} + d_{12}) \\ a_{11} + c_{12} - c_{11} - a_{12} \\ (b_{11}\text{-}d_{11}\text{-}b_{12} + d_{12}) \\ -(a_{11} + c_{12} - c_{11} - a_{12}) \\ [0] \\ \vdots \\ [0] \end{pmatrix}$$

and

$$
\begin{pmatrix}
a_{12}+c_{13}-c_{12}-a_{13}\\
0\\
-(a_{12}+c_{13}-c_{12}-a_{13})\\
0\\
-(a_{11}-c_{11}+c_{13}-a_{13})\\
0\\
a_{11}-c_{11}+c_{13}-a_{13}\\
0\\
a_{11}+c_{12}-c_{11}-a_{12}\\
0\\
-(a_{11}+c_{12}-c_{11}-a_{12})\\
0\\
[0]\\
\vdots\\
[0]
\end{pmatrix},
\begin{pmatrix}
b_{11}-d_{11}-b_{13}+d_{13}\\
-(a_{11}+c_{12}-c_{11}-a_{12})\\
-(b_{11}-d_{11}-b_{13}+d_{13})\\
a_{11}+c_{12}-c_{11}-a_{12}\\
-(b_{11}-d_{11}-b_{13}+d_{13})\\
0\\
b_{11}-d_{11}-b_{13}+d_{13}\\
0\\
0\\
a_{11}+c_{12}-c_{11}-a_{12}\\
0\\
-(a_{11}+c_{12}-c_{11}-a_{12})\\
[0]\\
\vdots\\
[0]
\end{pmatrix}, \ldots,
\begin{pmatrix}
a_{12}+c_{1n}-c_{12}-a_{1n}\\
0\\
-(a_{12}+c_{1n}-c_{12}-a_{1n})\\
0\\
-(a_{11}-c_{11}+c_{1n}-a_{1n})\\
0\\
a_{11}-c_{11}+c_{1n}-a_{1n}\\
0\\
[0]\\
\vdots\\
[0]\\
a_{11}+c_{12}-c_{11}-a_{12}\\
0\\
-(a_{11}+c_{12}-c_{11}-a_{12})\\
0
\end{pmatrix},
\begin{pmatrix}
b_{11}-d_{11}-b_{1n}+d_{1n}\\
-(a_{11}+c_{12}-c_{11}-a_{12})\\
-(b_{11}-d_{11}-b_{1n}+d_{1n})\\
a_{11}+c_{12}-c_{11}-a_{12}\\
-(b_{11}-d_{11}-b_{1n}+d_{1n})\\
0\\
b_{11}-d_{11}-b_{1n}+d_{1n}\\
0\\
[0]\\
\vdots\\
[0]\\
0\\
a_{11}+c_{12}-c_{11}-a_{12}\\
0\\
-(a_{11}+c_{12}-c_{11}-a_{12})
\end{pmatrix}
$$

Proof: We will check first that these vectors belong to the nullspace of the matrix. When we multiply the first vector with the matrix, we obtain a vector having in the first row

$(a_{11}-c_{11})(b_{11}-d_{11}-b_{12}+d_{12})+(b_{11}-d_{11})(a_{11}-c_{11}-a_{12}+c_{12})-(a_{12}-c_{12})(b_{11}-d_{11}-b_{12}+d_{12})-$
$(b_{12}-d_{12})(a_{11}-c_{11}-a_{12}+c_{12})$
$=(b_{11}-d_{11}-b_{12}+d_{12})[(a_{11}-c_{11})-(a_{12}-c_{12})]-\{(a_{11}-c_{11}-a_{12}+c_{12})[(b_{11}-d_{11})-(b_{12}-d_{12})]\}=0$

Since we know that

$$(a_{11}-c_{11})=-\left(e_{11}-g_{11}\right),\ (b_{11}-d_{11})=-\left(f_{11}-h_{11}\right).$$

we obtain zero in the second row of the vector. Since the third and fourth rows of the squares are complementary to the first two rows, we deduce that the third and fourth rows of the vector are also zero. Now, the fifth entry of the vector is

$(a_{21}-c_{21})(b_{11}-d_{11}-b_{12}+d_{12})+(b_{21}-d_{21})(a_{11}-c_{11}-a_{12}+c_{12})-$
$(a_{22}-c_{22})(b_{11}-d_{11}-b_{12}+d_{12})-(b_{22}-d_{22})(a_{11}-c_{11}-a_{12}+c_{12})=$
$(b_{11}-d_{11}-b_{12}+d_{12})[(a_{21}-c_{21})-(a_{22}-c_{22})]-\{(a_{11}-c_{11}-a_{12}+c_{12})[(b_{21}-d_{21})-(b_{22}-d_{22})]\}$

We use the following relations according to our assumption

$$a_{22}=a_{12}+a_{21}-a_{11},b_{22}=b_{12}+b_{21}-b_{11},$$
$$c_{22}=c_{12}+c_{21}-c_{11},d_{22}=d_{12}+d_{21}-d_{11}.$$

and obtain

$(b_{11}-d_{11}-b_{12}+d_{12})[(a_{21}-c_{21})-(a_{12}+a_{21}-a_{11}-c_{12}-c_{21}+c_{11})]$
$-\{(a_{11}-c_{11}-a_{12}+c_{12})[(b_{21}-d_{21})-(b_{12}+b_{21}-b_{11}-d_{12}-d_{21}+d_{11})]\}$
$=(b_{11}-d_{11}-b_{12}+d_{12})[-(a_{12}-a_{11}-c_{12}+c_{11})]-(a_{11}-c_{11}-a_{12}+c_{12})[-(b_{12}-b_{11}-d_{12}+d_{11})]=0$

We continue checking all rows until we reach the last entry, which is

$$(a_{n1} - c_{n1})(b_{11} - d_{11} - b_{12} + d_{12}) + (b_{n1} - d_{n1})(a_{11} - c_{11} - a_{12} + c_{12}) -$$
$$(a_{n2} - c_{n2})(b_{11} - d_{11} - b_{12} + d_{12}) - (b_{n2} - d_{n2})(a_{11} - c_{11} - a_{12} + c_{12}) =$$
$$(b_{11} - d_{11} - b_{12} + d_{12})[(a_{n1} - c_{n1}) - (a_{n2} - c_{n2})] - (a_{11} - c_{11} - a_{12} + c_{12})[(b_{n1} - d_{n1}) - (b_{n2} - d_{n2})]$$

We use

$$a_{n2} = a_{12} + a_{n1} - a_{11}, b_{n2} = b_{12} + b_{n1} - b_{11},$$
$$c_{n2} = c_{12} + c_{n1} - c_{11}, d_{n2} = d_{12} + d_{n1} - d_{11}.$$

in order to obtain this value of the entry

$$(b_{11} - d_{11} - b_{12} + d_{12})[(a_{n1} - c_{n1}) - (a_{12} + a_{n1} - a_{11} - c_{12} - c_{n1} + c_{11})]$$
$$-\{(a_{11} - c_{11} - a_{12} + c_{12})[(b_{n1} - d_{n1}) - (b_{12} + b_{n1} - b_{11} - d_{12} - d_{n1} + d_{11})]\}$$
$$= (b_{11} - d_{11} - b_{12} + d_{12})[-(a_{12} - a_{11} - c_{12} + c_{11})] - (a_{11} - c_{11} - a_{12} + c_{12})[-(b_{12} - b_{11} - d_{12} + d_{11})] = 0$$

Hence, we finished checking the first vector.

Now, we turn our attention to the second vector. When we multiply the matrix with it, we obtain in the first entry.

$$(a_{11} - c_{11})(a_{12} - c_{12} - a_{13} + c_{13}) - (a_{12} - c_{12})(a_{11} - c_{11} - a_{13} + c_{13}) + (a_{13} - c_{13})(a_{11} - c_{11} - a_{12} + c_{12}) =$$
$$(a_{11} - c_{11})[(a_{12} - c_{12}) - (a_{13} - c_{13})] - (a_{12} - c_{12})[(a_{11} - c_{11}) - (a_{13} - c_{13})] + (a_{13} - c_{13})[(a_{11} - c_{11}) - (a_{12} - c_{12})]$$
$$= (a_{11} - c_{11})(a_{12} - c_{12}) - (a_{11} - c_{11})(a_{13} - c_{13}) - (a_{12} - c_{12})(a_{11} - c_{11}) + (a_{12} - c_{12})(a_{13} - c_{13}) + (a_{13} - c_{13})(a_{11} - c_{11})$$
$$- (a_{13} - c_{13})(a_{12} - c_{12}) = 0$$

Using the relations

$$(a_{11} - c_{11}) = -(e_{11} - g_{11})$$
$$(b_{11} - d_{11}) = -(f_{11} - h_{11})$$

we deduce that the second entry is also zero. In a similar manner we can deal with the third and fourth entries. The fifth entry will be

$$(a_{21} - c_{21})(a_{12} - c_{12} - a_{13} + c_{13}) - (a_{22} - c_{22})(a_{11} - c_{11} - a_{13} + c_{13}) + (a_{23} - c_{23})(a_{11} - c_{11} - a_{12} + c_{12})$$

We use the relations

$$a_{22} = a_{12} + a_{21} - a_{11}, c_{22} = c_{12} + c_{21} - c_{11}$$
$$a_{23} = a_{13} + a_{21} - a_{11}, c_{23} = c_{13} + c_{21} - c_{11}$$

to obtain for the fifth entry.

$$= (a_{21} - c_{21})[(a_{12} - c_{12}) - (a_{13} - c_{13})] - (a_{12} + a_{21} - a_{11} - c_{12} - c_{21} + c_{11})[(a_{11} - c_{11}) - (a_{13} - c_{13})]$$
$$+ (a_{13} + a_{21} - a_{11} - c_{13} - c_{21} + c_{11})[(a_{11} - c_{11}) - (a_{12} - c_{12})]$$
$$= (a_{21} - c_{21})(a_{12} - c_{12}) - (a_{21} - c_{21})(a_{13} - c_{13}) - (a_{21} - c_{21})(a_{11} - c_{11}) + (a_{21} - c_{21})(a_{13} - c_{13})$$
$$- (a_{12} - c_{12})(a_{11} - c_{11}) + (a_{12} - c_{12})(a_{13} - c_{13}) + (a_{11} - c_{11})^2 + (a_{11} - c_{11})(a_{13} - c_{13}) + (a_{13} - c_{13})(a_{11} - c_{11})$$
$$- (a_{13} - c_{13})(a_{12} - c_{12}) + (a_{21} - c_{21})(a_{11} - c_{11}) - (a_{21} - c_{21})(a_{12} - c_{12}) - (a_{11} - c_{11})^2 + (a_{12} - c_{12})(a_{11} - c_{11}) = 0$$

We continue checking the entries until we reach the last entry, which is

$$(a_{n1} - c_{n1})(a_{12} - c_{12} - a_{13} + c_{13}) - (a_{n2} - c_{n2})(a_{11} - c_{11} - a_{13} + c_{13}) + (a_{n3} - c_{n3})(a_{11} - c_{11} - a_{12} + c_{12})$$

Using the relations

$$a_{n2} = a_{12} + a_{n1} - a_{11}, c_{n2} = c_{12} + c_{n1} - c_{11}$$
$$a_{n3} = a_{13} + a_{n1} - a_{11}, c_{n3} = c_{13} + c_{n1} - c_{11}$$

we get

$$= (a_{n1} - c_{n1})[(a_{12} - c_{12}) - (a_{13} - c_{13})] - (a_{12} + a_{n1} - a_{11} - c_{12} - c_{n1} + c_{11})[(a_{11} - c_{11}) - (a_{13} - c_{13})]$$
$$+ (a_{13} + a_{n1} - a_{11} - c_{13} - c_{n1} + c_{11})[(a_{11} - c_{11}) - (a_{12} - c_{12})]$$
$$= (a_{n1} - c_{n1})(a_{12} - c_{12}) - (a_{n1} - c_{n1})(a_{13} - c_{13}) - (a_{n1} - c_{n1})(a_{11} - c_{11}) + (a_{n1} - c_{n1})(a_{13} - c_{13})$$
$$- (a_{12} - c_{12})(a_{11} - c_{11}) + (a_{12} - c_{12})(a_{13} - c_{13}) + (a_{11} - c_{11})^2 + (a_{11} - c_{11})(a_{13} - c_{13}) +$$
$$(a_{13} - c_{13})(a_{11} - c_{11}) - (a_{13} - c_{13})(a_{12} - c_{12}) + (a_{n1} - c_{n1})(a_{11} - c_{11}) - (a_{n1} - c_{n1})(a_{12} - c_{12}) -$$
$$(a_{11} - c_{11})^2 + (a_{12} - c_{12})(a_{11} - c_{11}) = 0$$

Hence, the second vector belongs to the nullspace of the $(4n \times 4n)$-matrix.

Similarly, we can check that all the other vectors are included in the nullspace of the $(4n \times 4n)$-matrix. We check the last vector (the $(2n - 3)$-th vector) belongs to the nullspace of the $(4n \times 4n)$-matrix. The first entry by matrix multiplication is:

$$(a_{11} - c_{11})(b_{11} - d_{11} - b_{1n} + d_{1n}) + (b_{11} - d_{11})(a_{11} - c_{11} - a_{12} + c_{12}) -$$
$$(a_{12} - c_{12})(b_{11} - d_{11} - b_{1n} + d_{1n}) - (b_{1n} - d_{1n})(a_{11} - c_{11} - a_{12} + c_{12}) =$$
$$(b_{11} - d_{11} - b_{1n} + d_{1n})[(a_{11} - c_{11}) - (a_{12} - c_{12})] - (a_{11} - c_{11} - a_{12} + c_{12})[(b_{11} - d_{11}) - (b_{1n} - d_{1n})] = 0$$

As before we deduce also that the second, third, and fourth entries are zero. The fifth entry is

$$(a_{21} - c_{21})(b_{11} - d_{11} - b_{1n} + d_{1n}) + (b_{21} - d_{21})(a_{11} - c_{11} - a_{12} + c_{12}) -$$
$$(a_{22} - c_{22})(b_{11} - d_{11} - b_{1n} + d_{1n}) - (b_{2n} - d_{2n})(a_{11} - c_{11} - a_{12} + c_{12}) =$$
$$(b_{11} - d_{11} - b_{1n} + d_{1n})[(a_{21} - c_{21}) - (a_{22} - c_{22})] - (a_{11} - c_{11} - a_{12} + c_{12})[(b_{21} - d_{21}) - (b_{2n} - d_{2n})] = (b_{11} - d_{11} - b_{1n} + d_{1n})$$
$$[(a_{21} - c_{21}) - (a_{12} + a_{21} - a_{11} - c_{12} - c_{21} + c_{11})]$$

We use the relations

$$a_{22} = a_{12} + a_{21} - a_{11}$$
$$b_{2n} = b_{1n} + b_{21} - b_{11}$$
$$c_{22} = c_{12} + c_{21} - c_{11}$$
$$d_{2n} = d_{1n} + d_{21} - d_{11}$$

Therefore, this entry is

$$(b_{11} - d_{11} - b_{1n} + d_{1n})[(a_{21} - c_{21}) - (a_{12} + a_{21} - a_{11} - c_{12} - c_{21} + c_{11})]$$
$$- \{(a_{11} - c_{11} - a_{12} + c_{12})[(b_{21} - d_{21}) - (b_{1n} + b_{21} - b_{11} - d_{1n} - d_{21} + d_{11})]\}$$
$$= (b_{11} - d_{11} - b_{1n} + d_{1n})[-(a_{12} - a_{11} - c_{12} + c_{11})] - \{(a_{11} - c_{11} - a_{12} + c_{12})[-(b_{1n} - b_{11} - d_{1n} + d_{11})]\} = 0$$

When we reach the $(2n - 3)$th entry, we find that it is

$$(a_{n1} - c_{n1})(b_{11} - d_{11} - b_{1n} + d_{1n}) + (b_{n1} - d_{n1})(a_{11} - c_{11} - a_{12} + c_{12})-$$
$$(a_{nn} - c_{nn})(b_{11} - d_{11} - b_{1n} + d_{1n}) - (b_{nn} - d_{nn})(a_{11} - c_{11} - a_{12} + c_{12}) =$$
$$(b_{11} - d_{11} - b_{1n} + d_{1n})[(a_{n1} - c_{n1}) - (a_{nn} - c_{nn})]-$$
$$\{(a_{11} - c_{11} - a_{12} + c_{12})[(b_{n1} - d_{n1}) - (b_{nn} - d_{nn})]\}$$

We use the relations

$$a_{nn} = a_{1n} + a_{n1} - a_{11}$$
$$b_{nn} = b_{1n} + b_{n1} - b_{11}$$
$$c_{nn} = c_{1n} + c_{n1} - c_{11}$$
$$d_{nn} = d_{1n} + d_{n1} - d_{11}$$

to prove that this entry is

$$(b_{11} - d_{11} - b_{1n} + d_{1n})[(a_{n1} - c_{n1}) - (a_{12} + a_{n1} - a_{11} - c_{12} - c_{n1} + c_{11})]$$
$$-\{(a_{11} - c_{11} - a_{12} + c_{12})[(b_{n1} - d_{n1}) - (b_{1n} + b_{n1} - b_{11} - d_{1n} - d_{n1} + d_{11})]\}$$
$$= (b_{11} - d_{11} - b_{1n} + d_{1n})[-(a_{12} - a_{11} - c_{12} + c_{11})] - (a_{11} - c_{11} - a_{12} + c_{12})[-(b_{1n} - b_{11} - d_{1n} + d_{11})] = 0$$

We prove now that the vectors are linearly independent. Let $k_1, k_2, k_3, \ldots, k_{2n-4}, k_{2n-3} \in R$ such that

$$k_1 \begin{pmatrix} b_{11} - d_{11} - b_{12} + d_{12} \\ -(a_{11} + c_{12} - c_{11} - a_{12}) \\ -(b_{11} - d_{11} - b_{12} + d_{12}) \\ (a_{11} + c_{12} - c_{11} - a_{12}) \\ -(b_{11} - d_{11} - b_{12} + d_{12}) \\ a_{11} + c_{12} - c_{11} - a_{12} \\ (b_{11} - d_{11} - b_{12} + d_{12}) \\ -(a_{11} + c_{12} - c_{11} - a_{12}) \\ 0 \\ 0 \\ 0 \\ 0 \\ \vdots \\ 0 \\ 0 \\ 0 \\ 0 \end{pmatrix} + k_2 \begin{pmatrix} a_{12} + c_{13} - c_{12} - a_{13} \\ 0 \\ -(a_{12} + c_{13} - c_{12} - a_{13}) \\ 0 \\ -(a_{11} - c_{11} + c_{13} - a_{13}) \\ 0 \\ a_{11} - c_{11} + c_{13} - a_{13} \\ 0 \\ a_{11} + c_{12} - c_{11} - a_{12} \\ 0 \\ -(a_{11} + c_{12} - c_{11} - a_{12}) \\ 0 \\ \vdots \\ 0 \\ 0 \\ 0 \\ 0 \end{pmatrix} + \ldots + k_{2n-3} \begin{pmatrix} b_{11} - d_{11} - b_{1n} + d_{1n} \\ -(a_{11} + c_{12} - c_{11} - a_{12}) \\ -(b_{11} - d_{11} - b_{1n} + d_{1n}) \\ a_{11} + c_{12} - c_{11} - a_{12} \\ -(b_{11} - d_{11} - b_{1n} + d_{1n}) \\ 0 \\ b_{11} - d_{11} - b_{1n} + d_{1n} \\ 0 \\ 0 \\ 0 \\ 0 \\ 0 \\ \vdots \\ 0 \\ a_{11} + c_{12} - c_{11} - a_{12} \\ 0 \\ -(a_{11} + c_{12} - c_{11} - a_{12}) \end{pmatrix} = \begin{pmatrix} 0 \\ 0 \\ 0 \\ 0 \\ 0 \\ 0 \\ 0 \\ 0 \\ 0 \\ 0 \\ 0 \\ 0 \\ \vdots \\ 0 \\ 0 \\ 0 \\ 0 \end{pmatrix}$$

This leads us to the following vector which is a zero vector.

$$\begin{pmatrix} k_1(b_{11} - d_{11} - b_{12} + d_{12}) + k_2(a_{12} + c_{13} - c_{12} - a_{13}) + k_3(b_{11} - d_{11} - b_{13} + d_{13}) + \ldots + \\ k_{2n-4}(a_{12} + c_{1n} - c_{12} - a_{1n}) + k_{2n-3}(b_{11} - d_{11} - b_{1n} + d_{1n}) \\ -k_1(a_{11} + c_{12} - c_{11} - a_{12}) - k_3(a_{11} + c_{12} - c_{11} - a_{12}) - k_{2n-3}(a_{11} + c_{12} - c_{11} - a_{12}) \\ -k_1(b_{11} - d_{11} - b_{12} + d_{12}) - k_2(a_{12} + c_{13} - c_{12} - a_{13}) - k_3(b_{11} - d_{11} - b_{13} + d_{13}) - \ldots - \\ k_{2n-4}(a_{12} + c_{1n} - c_{12} - a_{1n}) - k_{2n-3}(b_{11} - d_{11} - b_{14} + d_{1n}) \\ k_1(a_{11} + c_{12} - c_{11} - a_{12}) + k_3(a_{11} + c_{12} - c_{11} - a_{12}) + k_{2n-3}(a_{11} + c_{12} - c_{11} - a_{12}) \\ -k_1(b_{11} - d_{11} - b_{12} + d_{12}) - k_2(a_{11} - c_{11} + c_{13} - a_{13}) - k_3(b_{11} - d_{11} - b_{13} + d_{13}) - \ldots - \\ k_{2n-4}(a_{11} - c_{11} + c_{1n} - a_{1n}) - k_{2n-3}(b_{11} - d_{11} - b_{1n} + d_{1n}) \\ k_1(a_{11} + c_{12} - c_{11} - a_{12}) \\ k_1(b_{11} - d_{11} - b_{12} + d_{12}) + k_2(a_{11} - c_{11} + c_{13} - a_{13}) + k_3(b_{11} - d_{11} - b_{13} + d_{13}) + \ldots + \\ k_{2n-4}(a_{11} - c_{11} + c_{1n} - a_{1n}) + k_{2n-3}(b_{11} - d_{11} - b_{1n} + d_{1n}) \\ -k_1(a_{11} + c_{12} - c_{11} - a_{12}) \\ k_2(a_{11} + c_{12} - c_{11} - a_{12}) \\ k_3(a_{11} + c_{12} - c_{11} - a_{12}) \\ -k_2(a_{11} + c_{12} - c_{11} - a_{12}) \\ -k_3(a_{11} + c_{12} - c_{11} - a_{12}) \\ \vdots \\ k_{2n-4}(a_{11} + c_{12} - c_{11} - a_{12}) \\ k_{2n-3}(a_{11} + c_{12} - c_{11} - a_{12}) \\ -k_{2n-4}(a_{11} + c_{12} - c_{11} - a_{12}) \\ -k_{2n-3}(a_{11} + c_{12} - c_{11} - a_{12}) \end{pmatrix}$$

From the $(4n - 2)$-th row of this vector we obtain the equation

$$k_{2n-3}(a_{11} + c_{12} - c_{11} - a_{12}) = 0$$

According to our assumptions we must have $k_{2n-3} = 0$. Similarly, we obtain $k_{2n-4} = 0$ from the $(4n - 3)$-th row. We continue checking all the rows up to the tenth row, which looks like this

$$k_3(a_{11} + c_{12} - c_{11} - a_{12}) = 0$$

Hence, we conclude that $k_3 = 0$. From the ninth (res. eighth) row we obtain $k_2 = 0$ (res. $k_1 = 0$). Since all $k_1, k_2, k_3, \ldots, k_{2n-4}, k_{2n-3}$ are zero, we are done. $\square$

Author details

Saleem Al-Ashhab

Address all correspondence to: ahhab@aabu.edu.jo

Department of Mathematics, Al-albayt University, Mafraq, Jordan

References

[1] Ahmed M. Algebraic combinatorics of magic squares [Ph.D. thesis]. University of California, Davis: Mathematics Dep.; 2004

[2] Al-Amerie M. Msc. thesis, Al-Albayt University, supervised by S. Al-Ashhab. 2007

[3] Al-Ashhab S. Theory of Magic Squares. Royal Scientific Society; 2000

[4] Al-Ashhab S, Mueller W, Semi pandiagonal magic 4×4 squares. Results in Mathematics. 2003;**44**:25-28

[5] Andress W. Basic properties of pandiagonal magic squares. The American Mathematical Monthly. 1960;**67**(2):143-152

[6] Hendricks J. The determinant of a pandiagonal magic square of order 4 is zero. Journal of Recreational Mathematics. 1989;**21**:179-181

[7] Rosser B, Walker R. On the transformation group for diabolic magic squares of order four. Bulletin of the American Mathematical Society. 1938;**44**:416-420

[8] Kraitchik. La Mathématique des Jeux ou récréations mathématiques. p. 167

Chapter 5

Nature of Phyllotaxy and Topology of H-matrix

Ab. Hamid Ganie

Additional information is available at the end of the chapter

http://dx.doi.org/10.5772/intechopen.74676

Abstract

The main purpose of this chapter is to introduce a new type of regular matrix generated by Fibonacci numbers and we shall investigate its various topological properties. The concept of mathematical regularity in terms of Fibonacci numbers and phyllotaxy have been discussed.

Keywords: sequence spaces, infinite matrices, Fibonacci numbers, phyllotaxy
AMS Mathematical Subject Classification (2010); 46A45; 11B39d; 40C05

1. Preliminaries, background and notation

In several branches of analysis, for instance, the structural theory of topological vector spaces, Schauder basis theory, summability theory, and the theory of functions, the study of sequence spaces occupies a very prominent position. There is an ever-increasing interest in the theory of sequence spaces that has made remarkable advances in enveloping summability theory via unified techniques effecting matrix transformations from one sequence space into another.

Thus, we have several important applications of the theory of sequence spaces, and therefore, we attempt to present a survey on recent developments in sequence spaces and their different kinds of duals.

In many branches of science and engineering, we deal with different kinds of sequences and series, and when we deal with these, it is important to check their convergence. The use of infinite matrices is of great importance, we can bring even the bounded or divergent sequences and series in the domain of convergence. So we can say that the theory of sequence spaces and their matrix maps is the bigger scale to measure the convergence property. Summability can be roughly considered as the study of linear transformations on sequence spaces. The theory

© 2018 The Author(s). Licensee IntechOpen. This chapter is distributed under the terms of the Creative Commons Attribution License (http://creativecommons.org/licenses/by/3.0), which permits unrestricted use, distribution, and reproduction in any medium, provided the original work is properly cited. (cc) BY

IntechOpen

originated from the attempts of mathematicians to assign limits to divergent sequences. The classical summability theory deals with the generalization of the convergence of sequences or series of real or complex numbers. The idea is to assign a limit of some sort to divergent sequences or series by considering a transform of a sequence or series rather than the original sequence or series.

The earliest idea of summability theory was perhaps contained in a letter written by Leibnitz to C. Wolf (1713) in which he attributed the sum 1/2 to the oscillatory series $-1 + 1-1 + \ldots$. Frobenius in (1880) introduced the method of summability by arithmetic means, which was generalized by Cesàro in (1890) as the (C,K) method of summability. Toward the end of the nineteenth century, study of the general theory of sequences and transformations on them attracted mathematicians, who were chiefly motivated by problems such as those in summability theory, Fourier series, power series and system of equations with infinitely many variables.

Presenting some basic definitions and notations that are involved in the present work, the author proposes to give a brief resume of the hitherto obtained results against the background of which the main results studied in the present chapter suggest themselves.

2. Notations and symbols

Here, we state a few conventions which will be used throughout the chapter.

2.1. Symbols $\mathbb{N}$, $\mathbb{C}$, $\mathbb{R}$ and A

The symbols are denoted as follows:

$\mathbb{N}$: Set of non-negative integers.

$\mathbb{C}$: Set of complex numbers.

$\mathbb{R}$: Set of real numbers.

A: The infinite matrix (a_{nk}), $(n, k = 1, 2, \ldots)$.

2.2. Summation convention

By $\sum_{\alpha}^{\beta} f(n)$, we mean the sum of all values of $f(n)$ for which $\alpha \leq n \leq \beta$. In the case $\beta < \alpha$, then we take this to be zero.

Summations are over $0, 1, 2, \ldots$, when there is no indication to the contrary. If $(x_k) = (x_1, x_2, \ldots)$ is a sequence of terms, then, by $\sum_k x_k$ we mean $\sum_{k=1}^{\infty} x_k$ and we shall sometimes write as $\sum x_k$ incase where no possible confusion arises.

2.3. The spaces ω, l_∞, c, c_0, l_p

A sequence space is a set of scalar sequences (real or complex) which is closed under coordinate-wise addition and scalar multiplication. In other words, a sequence space is a linear subspace of the space ω of all complex sequences, that is,

$$\omega = \{ x = (x_k) : x_k \in \mathbb{R} \text{ or } \mathbb{C} \}.$$

The space l_∞: The space l_∞ of bounded sequences is defined by

$$\left\{ x = (x_k) : \sup_k |x_k| < \infty \right\}$$

The spaces c: The spaces c and c_0 of convergent and null sequences are given by

$$\left\{ x = (x_k) : \lim_k x_k = l, l \in \mathbb{C} \right\}$$

The space c_0: The space c_0 of all sequences converging to 0 is given by

$$\left\{ x = (x_k) : \lim_k x_k = 0 \right\}$$

The space l_p: The space l_p of absolutely p-summable sequences is defined by

$$\left\{ x = (x_k) : \sum_k |x_k|^p < \infty \right\}, (0 < p < \infty)$$

The spaces $l_\infty, c,$ and c_0 are Banach spaces with the norm,

$$\|x\|_\infty = \sup_k |x_k|$$

The space l_p is a Banach space with the norm,

$$\|x\|_p = \left(\sum_k |x_k|^p \right)^{\frac{1}{p}}, 1 \leq p < \infty$$

2.4. Cauchy sequence

A sequence $x = (x_k)$ is called a Cauchy sequence if and only if $|x^n - x^m| \to 0$ $(m, n \to \infty)$ that is for any $\epsilon > 0$, there exists $N = N(\epsilon)$ such that $|x^n - x^m| < \epsilon$ for all $n, m \geq N$. By $\mathfrak{C}$, we denote the space of all Cauchy sequences, that is,

$$\mathfrak{C} : \{x = (x_k) : |x^n - x^m| \to 0 \text{ as } n, m \to \infty\}$$

2.5. FK-space

A sequence space X is called an FK-space if it is a complete linear metric space with continuous coordinates $p_n : X \to \mathbb{C}$ defined by $p_n(x) = x_n$ for all $x \in X$ and every $n \in \mathbf{N}$ [1, 2].

2.6. BK-space

A BK-space is a normed FK-space, that is, a BK-space is a Banach space with continuous coordinates [3–6].

2.7. Fibonacci numbers

In the 1202 AD, Leonardo Fibonacci wrote in his book Liber Abaci of a simple numerical sequence that is the foundation for an incredible mathematical relationship behind phi. This sequence was known as early as the sixth century AD by Indian mathematicians, but it was Fibonacci who introduced it to the west after his travels throughout the Mediterranean world and North Africa. He is also known as Leonardo Bonacci, as his name is derived in Italian from words meaning son of (the) Bonacci.

The Fibonacci numbers have been introduced [7–14]. The Fibonacci numbers are the sequence of numbers $\{f_n\}, n \in \mathbb{N}$ defined by recurrence relations

$$f_0 = 0, f_1 = 1 \text{ and } f_n = f_{n-1} + f_{n-2}; n \geq 2$$

First derived from the famous rabbit problem of 1228, the Fibonacci numbers were originally used to represent the number of pairs of rabbits born of one pair in a certain population. Let us assume that a pair of rabbits is introduced into a certain place in the first month of the year. This pair of rabbits will produce one pair of offspring every month, and every pair of rabbits will begin to reproduce exactly 2 months after being born. No rabbit ever dies, and every pair of rabbits will reproduce perfectly on schedule.

Month	Pairs	Number of pairs of adults (A)	Number of pairs of babies (B)	Total pairs
January 1		1	0	1
February 1		1	1	2
March 1		2	1	3
April 1		3	2	5
May 1		5	3	8
June 1		8	5	13
July 1		13	8	21
August 1		21	13	34
September 1		34	21	55
October 1		55	34	89
November 1		89	55	144
December 1		144	89	233
January 1		233	144	377

The number of pairs of mature rabbits living each month determines the Fibonacci sequence (column 1): 1, 1, 2, 3, 5, 8, 13, 21, 34, 55, 89, 144, 233, 377, ...

So, in the first month, we have only the first pair of rabbits. Likewise, in the second month, we again have only our initial pair of rabbits. However, by the third month, the pair will give birth to another pair of rabbits, and there will now be two pairs. Continuing on, we find that in month 4, we will have 3 pairs, then 5 pairs in month 5, then 8, 13, 21, 34, ..., etc., continuing in this manner. It is quite apparent that this sequence directly corresponds with the Fibonacci sequence introduced above, and indeed, this is the first problem ever associated with the now-famous numbers.

Fibonacci numbers have many interesting properties and applications in arts, sciences and architecture. Also, following [7], some basic properties are as follows

$$\sum_{k=0}^{n} f_k = f_{n+2} - 1; n \in \mathbf{N},$$

and

$$\sum_{k=0}^{n} f_k^2 = f_n f_{n+1}; n \in \mathbf{N}$$

Everything in Nature is subordinated to stringent mathematical laws. Prove to be that leaf's disposition on plant's stems also has stringent mathematical regularity and this phenomenon is called phyllotaxis in botany. An essence of phyllotaxis consists in a spiral disposition of leaves on plant's stems of trees, petals in flower baskets, seeds in pine cone and sunflower head, etc.

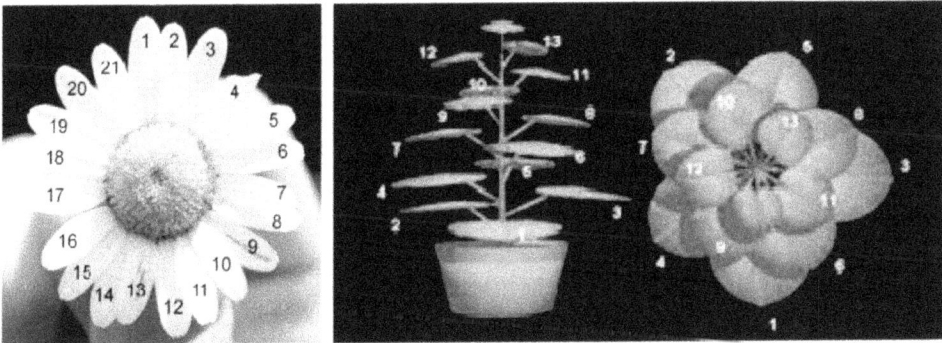

This phenomenon, known already to Kepler, was a subject of discussion of many scientists, including Leonardo da Vinci, Turing, Veil, and so on. In phyllotaxis phenomenon, more complex concepts of symmetry, in particular, a concept of helical symmetry, are used. The phyllotaxis phenomenon reveals itself especially brightly in inflorescences and densely packed botanical structures such as pine cones, pineapples, cacti, heads of sunflower and cauliflower, and many other objects [11].

On the surfaces of such objects, their bio-organs (seeds on the disks of sunflower heads and pine cones, etc.) are placed in the form of the left-twisted and right-twisted spirals. For such phyllotaxis objects, it is used usually the number ratios of the left-hand and right-hand spirals observed on the surface of the phyllotaxis objects. Botanists proved that these ratios are equal to the ratios of the adjacent Fibonacci numbers, that is,

$$\frac{f_{i+1}}{f_i} : \frac{2}{1}, \frac{3}{2}, \frac{5}{3}, \frac{8}{5}, \frac{13}{8}, \dots = \frac{1+\sqrt{5}}{2}$$

By using hyperbolic Fibonacci functions, he had developed an original geometric theory of phyllotaxis and explained why Fibonacci spirals arise on the surface of the phyllotaxis objects namely, pine cones, cacti, pine apple, heads of sunflower, and so on, in process of their growths. Bodnar's geometry [15] confirms that these functions are 'natural' functions of the nature, which show their value in the botanic phenomenon of phyllotaxis. This fact allows us to assert that these functions can be attributed to the class of fundamental mathematical discoveries of contemporary science because they reflect natural phenomena, in particular, phyllotaxis phenomenon.

From above discussion, it gave us motivation to see the behavior of the infinite matrices generated by Fibonacci numbers.

In the present chapter, we have introduced a new type of matrix $H = (h_{nk}^u) \, n, k \in \mathbb{N}$ by using Fibonacci numbers f_n and we call it as H-matrix generated by Fibonacci numbers f_n and introduce some new sequence spaces related to matrix domain of H in the sequence spaces l_p, l_∞, c and c_0, where $1 \leq p < \infty$.

2.8. The space $r^q(u, p)$

Sheikh and Ganie [16] introduced the Riesz sequence space $r^q(u, p)$ and studied its various topological properties where $u = (u_k)$ is a sequence such that $u_k \neq 0$ for all $k \in \mathbb{N}$ and (q_k) the sequence of positive numbers and

$$Q_n = \sum_{k=0}^{n} q_k, \forall n \in \mathbb{N}$$

Then, the matrix $R_u^q = (r_{nk}^q)$ of the Riesz mean (R_u, q_n) is given by

$$r_{nk}^q = \begin{cases} \dfrac{u_k q_k}{Q_n} & \text{if } 0 \leq k \leq n, \\ 0, & \text{if } k > n. \end{cases}$$

The Riesz mean (R_u, q_n) is regular if and only if $Q_n \to \infty$ as $n \to \infty$.

3. H-matrix generated by Fibonacci numbers

Let X and Y be two subsets of ω. Let $A = (a_{nk})$ be an infinite matrix of real or complex numbers a_{nk}, where $n, k \in \mathbb{N}$. Then, the matrix A defines the A-transformation from X into Y, if for every sequence $x = (x_k) \in X$ the sequence $Ax = \{(Ax)_n\}$, the A-transform of x exists and is in Y where

$$(Ax)_n = \sum_k a_{nk}x_k.$$

For simplicity in notation, here and in what follows, the summation without limits runs from 0 to ∞. By (X, Y), we denote the class of all such matrices. A sequence x is said to be A-summable to l if Ax converges to l which is called as the A-limit of x.

For a sequence space X, the matrix domain X_A of an infinite matrix A is defined as

$$X_A = \{x = (x_k) \in \omega : Ax \in X\}, \tag{1}$$

which is a sequence space.

An infinite matrix $A = (a_{nk})$ is said to be regular if and only if the following conditions (or Toplitz conditions) hold [17–19]:

i. $\lim\limits_{n \to \infty} \sum\limits_{k=0}^{\infty} a_{nk} = 1,$

ii. $\lim\limits_{n \to \infty} a_{nk} = 0, \quad (k = 0, 1, 2, \ldots),$

iii. $\sum\limits_{k=0}^{\infty} |a_{nk}| < M, \quad (M > 0, j = 0, 1, 2, \ldots).$

In the present paper, we introduce H-matrix with $H = (h_{nk}^u) \, n, k \in \mathbb{N}$ as follows:

$$h_{nk}^u = \begin{cases} \dfrac{u_k f_k^{\,2}}{f_n f_{n+1}} & \text{if } 0 \le k \le n, \\ 0, & \text{if } k > n. \end{cases}$$

Thus, for $u_k = 1$ and for all $k \in \mathbb{N}$, we have

$$H = \begin{pmatrix} 1 & 0 & 0 & 0 & 0 & \cdots \\ 1/2 & 1/2 & 0 & 0 & 0 & \cdots \\ 1/6 & 1/6 & 4/6 & 0 & 0 & \cdots \\ 1/15 & 1/15 & 4/15 & 9/15 & 0 & \cdots \\ \vdots & \vdots & \vdots & \vdots & \ddots & \vdots \end{pmatrix}.$$

It is obvious that the matrix H is a triangle, that is, $h_{nn}^u \ne 0$ and $h_{nk}^u = 0$ for $k > n$ and for all $n \in \mathbb{N}$. Also, since it satisfies the conditions of Toeplitz matrix and hence it is regular matrix.

Note that if we take $q_k = f_k^2$, then the matrix H is special case of the matrix R_u^q, where

$$Q_n = \sum_{k=0}^{n} f_k^2 = f_n f_{n+1},$$

introduced by Sheikh and Ganie [16].

The approach of constructing a new sequence space by means of matrix domain of a particular limitation method has been studied by several authors [17–26].

Throughout the text of the chapter, X denotes any of the spaces l_∞, c, c_0 and l_p ($1 \leq p < \infty$). Then, the Fibonacci sequence space $X(H)$ is defined by

$$X(H) = \{x = (x_k) \in \omega : y = (y_k) \in X\},$$

where the sequence $y = (y_k)$ is the H-transform of the sequence $x = (x_k)$ and is given by

$$y_k = H_k(x) = \frac{1}{f_k f_{k+1}} \sum_{i=0}^{k} f_i^2 u_i x_i \text{ for all } k \in \mathbb{N}. \tag{2}$$

With the definition of matrix domain given by Eq. (1), we can redefine the space $X(H)$ as the matrix domain of the triangle H in the space X, that is,

$$X(H) = X_H.$$

Theorem 1: The space $X(H)$ is a BK-space with the norm given by

$$\|x\| = |H(x)\|_X = \|y\|_X = \begin{cases} \left[\sum_{k=0}^{\infty} |y_k|^p\right]^{\frac{1}{p}} & \text{for for } X \in \{l_p\}. \\ \sup_k y_k & \text{for } X \in \{l_\infty, c, c_0\}. \end{cases} \tag{3}$$

Proof: Since the matrix $H = (h_{nk}^u)$ is a triangle, that is, $h_{nn}^u \neq 0$ and $h_{nk}^u = 0$ for $k > n$ for all n. We have the result by Eq. (3) and Theorem 4.3.2 of Wilansky [6] gives the fact that $X(H)$ is a BK-space.◊

Theorem 2: The space $X(H)$ is isometrically isomorphic to the space X.

Proof: To prove the result, we should show the linear bijection between the spaces $X(H)$ and X. For that, consider the transformation T from $X(H)$ to X by $x \to y = Tx$. Then, the linearity of T follows from Eq. (2). Further, we see that $x = 0$ whenever $Tx = 0$ and consequently T is injective.

Moreover, let $y = (y_k) \in X$ be given and define the sequence $x = (x_k)$ by

$$x_k = \frac{f_{k+1}}{u_k f_k} y_k - \frac{f_{k-1}}{u_k f_k} y_{k-1}; k \in \mathbb{N}. \tag{4}$$

Then, by using (2) and (4), we have for every $k \in \mathbb{N}$ that

$$H(x) = \frac{1}{f_k f_{k+1}} \sum_{i=0}^{k} f_i^2 u_i x_i$$

$$= \frac{1}{f_k f_{k+1}} \sum_{i=0}^{k} f_i (f_{i+1} y_i - f_{i-1} y_{i-1})$$

$$= y_k.$$

This shows that $H(x) = y$ and since $y \in X$, we conclude that $H(x) \in X$. Thus, we deduce that $x \in X(H)$ and $Tx = y$. Hence, T is surjective.

Furthermore, for any $x \in X(H)$, we have by (3) that

$$\|T(x)\| = \|y\| = \|H(x)\|_X = \|x\|_X$$

which shows that T is norm preserving. Hence, T is isometry. Consequently, the spaces $X(H)$ and X are isometrically isomorphic. Hence, the proof of the Theorem is complete.◊

Theorem 3: Let $\{f_i\}$ be Fibonacci number sequences. Then, we have

$$\sup_i \left(f_i^2 \sum_{j=i}^{\infty} \frac{1}{f_j f_{j+1}} \right) < \infty.$$

Proof: We have,

$$\sum_{k=n}^{\infty} \left(\frac{1}{f_k} - \frac{1}{f_{k+1}} \right) = \frac{1}{f_n}$$

This gives,

$$1 = f_n \sum_{k=n}^{\infty} \left(\frac{1}{f_k} - \frac{1}{f_{k+1}} \right)$$

$$= f_n^2 \frac{1}{f_n} \sum_{k=n}^{\infty} \left(\frac{f_{k+1} - f_k}{f_k f_{k+1}} \right)$$

$$= f_n^2 \frac{1}{f_n} \sum_{k=n}^{\infty} \left(\frac{f_{k-1}}{f_k f_{k+1}} \right)$$

$$\geq f_n^2 \frac{f_{n-1}}{f_n} \sum_{k=n}^{\infty} \left(\frac{1}{f_k f_{k+1}} \right)$$

and the conclusion follows because $f_n f_{n-1}$ is bounded since it converges to $\frac{\sqrt{5}+1}{2}$.◊

Theorem 4: $X \subset X(H)$ holds.

Proof: It is obvious that $c_0 \subset c_0(H)$ and $c \subset c(H)$, since the matrix H is regular matrix. Now, let $x \in l_\infty$. Then, there is a constant $K > 0$ such that $|x_j| < \frac{K}{|u_j|}$ for all $j \in \mathbf{N}$. Thus, we have for every $i \in \mathbf{N}$ that

$$|II_i(x)| \leq \frac{1}{f_i f_{i+1}} \sum_{j=0}^{i} f_j^2 |u_j x_j|$$

$$\leq \frac{K}{f_i f_{i+1}} \sum_{j=0}^{i} f_j^2 = K$$

which shows that $H(x) \in l_\infty$. Therefore, we deduce that $x \in l_\infty$ implies $x \in l_\infty(H)$.

We now consider the case $1 \leq p < \infty$. We only consider the case $1 < p < \infty$ and by similar argument will follow for $p = 1$. So, let $x \in l_p$. Then, for every $i \in \mathbf{N}$ and by Holder's inequality, we have

$$|H_i(x)|^p \leq \left(\sum_{j=0}^{i} \frac{f_j^2}{f_i f_{i+1}} |u_j x_j| \right)^p$$

$$\leq \left(\sum_{j=0}^{i} \frac{f_j^2}{f_i f_{i+1}} |u_j x_j| \right)^p \left(\sum_{j=0}^{i} \frac{f_j^2}{f_i f_{i+1}} \right)^{p-1}$$

$$= \frac{1}{f_i f_{i+1}} \sum_{j=0}^{i} f_j^2 |u_j x_j|^p.$$

Hence, we have

$$\sum_{i=0}^{\infty} |H_i(x)|^p \leq \sum_{i=0}^{\infty} \frac{1}{f_i f_{i+1}} \sum_{j=0}^{i} f_j^2 |u_j x_j|^p$$

$$= \sum_{i=0}^{\infty} |x_j|^p |u_j|^p f_j^2 \sum_{i=j}^{\infty} \frac{1}{f_i f_{i+1}}.$$

Hence, the right-hand side of above inequality can be made arbitrary small, since, $\sup_j \left(f_j^2 \sum_{i=j}^{\infty} \frac{1}{f_i f_{i+1}} \right) < \infty$ by Theorem 3 (above) and $x \in l_p$. This shows that $x \in l_p(H)$. This completes the proof of the theorem. ◊

Acknowledgements

The author would like to express his sincere thanks for the refree(s) for the kind remarks that improved the presentation of the chapter.

Author details

Ab. Hamid Ganie

Address all correspondence to: ashamidg@rediffmail.com

Department of Applied Science and Humanities, SSM College of Engineering and Technology, Pattan, Jammu and Kashmir, India

References

[1] Altay B, Başar F. On the paranormed Riesz sequence spaces of non-absolute type. Southeast Asian Bulletin of Mathematics. 2002;**26**(5):701-715

[2] Mursaleen M, Ganie AH, Sheikh NA. New type of generalized difference sequence space of non-absolute type and some matrix transformations. Filomat. 2014;**28**(7):1381-1392

[3] Boos J. Classical and Modern Methods in Summability. Oxford, UK: Oxford University Press; 2000

[4] Maddox IJ. Elements of Functional Analysis. 2nd ed. Cambridge: University Press; 1988

[5] Toeplitz O. Uber allegemeine Lineare mittelbildungen. Prace Matematyczno Fizyczne. 1991;**22**:113-119

[6] Wilansky A. Summability through functional analysis. North-Holland Mathematics Studies. 1984;**85**

[7] Koshy T. Fibonacci and Lucas Numbers with Applications. Wiley; 2001

[8] Alexey S, Samuil A. Hyperbolic Fibonacci and Lucas functions, golden Fibonacci goniometry, Bodnars geometry, and Hilberts fourth problem. Applied Mathematics. 2011;**2**: 181-188

[9] Kalman D, Mena R. The Fibonacci numbers-exposed. Mathematics Magazine. 2003;**76**(3)

[10] Kirschenhofer P, Prodinger H, Tichy RF. Fibonacci numbers of graphs: III: Planted plane trees. In: Fibonacci Numbers and their Applications. Dordrecht: D. Reidel; 1986. pp. 105-120

[11] de Malafosse B. Properties of some sets of sequences and application to the spaces of bounded difference sequences of order μ. Hokkaido Mathematical Journal. 2002;**31**(2): 283-299

[12] Stakhov A. The general principle of the golden section and its applications in mathematics, science and engineering. Chaos, Solutions and Fractals. 2005;**26**:263-289

[13] Ming KX. Generalized Fibonacci sequence. Higher Mathematics. 2007;**10**(1):60-64

[14] Zhang JP. A class of generalized Fibonacci sequence and its application. Quanzhou Normal University (Natural Science). 2005;**23**(2):10-13

[15] Bodnar OY. The Golden Section and Non-Euclidean Geometry in Nature and Art. Lvov: Svit; 1994. (In Russian)

[16] Sheikh NA, Ganie AH, A new paranormed sequence space and some matrix transformations. Acta Mathematica Academiae Paedagoglace Nyíregyháziensis, 2012;**28**:47-58

[17] Petersen GM. Regular Matrix Transformations. London: McGraw-Hill; 1966

[18] Wang CS. On Norlund sequence spaces. Tamkang Journal of Mathematics. 1978;**9**:269-274

[19] Altay B, Başar F. On the space of sequences of p-bounded variation and related matrix mappings. Ukrainian Mathematical Journal. 2003;**55**(1):136-147

[20] Choudhary B, Mishra SK. On Kothe Toeplitz duals of certain sequence spaces and matrix transformations. Indian Journal of Pure and Applied Mathematics. 1993;**24**:291-301

[21] Ganie AH, Ahmad M, Sheikh NA, Jalal T. New type of Riesz sequence space of non-absolute type. Journal of Applied and Computational Mathematics. 2016;**5**:280

[22] Kizmaz H. On certain sequence spaces. Canadian Mathematical Bulletin. 1981;**24**(2): 169-175

[23] Maddox IJ. Paranormed sequence spaces generated by infinite matrices. Proceedings of the Cambridge Philosophical Society. 1968;**64**:335-340

[24] Metin B, Mahpeyker O. On the Riesz difference sequence space. Rendiconti del Circolo Matematico di Palermo. 2008;**57**:377-389

[25] Ng P-N, Lee P-Y. Cesáro sequences spaces of non-absolute type. Commentationes Mathematicae. Prace Matematyczne. 1978;**20**:429-433

[26] Lorentz GG, Zeller K. Summation of sequences and summation of series. Proceedings of American Mathematical Society. 1964;**15**:743-746

www.ingramcontent.com/pod-product-compliance
Lightning Source LLC
Chambersburg PA
CBHW081235190326

41458CB00016B/5787